OXFORD MEDICAL PUBLIC

ORAL MEDICIN

# ORAL MEDICINE

## THIRD EDITION

●

W. R. TYLDESLEY

*Dean of Dental Studies*
*University of Liverpool*

Oxford   New York   Tokyo
OXFORD UNIVERSITY PRESS
1989

Oxford University Press, Walton Street, Oxford OX2 6DP

Oxford   New York   Toronto
Delhi   Bombay   Calcutta   Madras   Karachi
Petaling Jaya   Singapore   Hong Kong   Tokyo
Nairobi   Dar es Salaam   Cape Town
Melbourne   Auckland

and associated companies in
Berlin   Ibadan

Oxford is a trade mark of Oxford University Press

Published in the United States
by Oxford University Press, New York

First edition 1974
Second edition 1981
Third edition 1989

British Library Cataloguing in Publication Data

Tyldesley, W. R. (William Randolph, 1927–   )
Oral medicine. – 3rd ed
1. Man. Mouth. Diseases
I. Title
616.3'1
ISBN 0–19–261896–2
0–19–261895–4 (Pbk)

Library of Congress Cataloging in Publication Data

Tyldesley, William R.
Oral medicine.
(Oxford medical publications)
Bibliography   Includes index.
1. Oral medicine.   I. Title.   II. Series. [DNLM:
1. Dentistry. WU 100T9810]
RC815.T9 1989      616.3'1      89–8643
ISBN 0–19–261896–2
ISBN 0–19–261895–4 (pbk.)


Set by Latimer Trend & Company Ltd, Plymouth
Printed in Great Britain
by Bookcraft (Bath) Ltd
Midsomer Norton, Avon

# PREFACE TO THIRD EDITION

THE purpose of this edition remains the same as that of the previous two: to provide an up-to-date survey of the field of oral medicine as a help for undergraduates and as a guide for practitioners. The scope of the book has been somewhat expanded—some subjects have been introduced for the first time and other sections amplified.

Two issues have been decided as a matter of policy. The first is in relation to references. It is a frequently heard complaint of reviewers that books of this kind are incomplete without original reference lists. The author differs from this view. A number of fully referenced review texts are available for the research worker or advanced student; the great majority of the readership for which the present book is aimed do not have the time or facility to follow up original references. It is, indeed, the function of this book to present a review of such information in a readily available and self-contained format. The second is in regard to illustrations. It is generally accepted that black and white illustrations of oral mucosal lesions are, at best, only partially successful. It has, none the less, been decided to increase their number to some extent and it is hoped that the illustrations in the present book are helpful within their limitations. Clearly, the cost of illustrating a book of this kind in colour would increase its price in a quite undesirable way. A number of colour atlases of oral medicine have been produced (by the author and by others) which present far more accurate representations of oral lesions than is possible in monochrome reproduction.

The final sentence of the preface to the previous edition still applies: it is hoped that the present version of this book will accurately reflect current ideas and attitudes in a rapidly changing subject.

*Liverpool*                                                                                          W.R.T.
November 1988

# ACKNOWLEDGEMENTS

JUST a little of the original material used in the *British Dental Journal* series may be recognized in the present edition; the Editor of the Journal has kindly agreed to its further use. Another undergraduate textbook produced by the author (*Oral Diagnosis*) has appeared in two English as well as in some foreign language editions. It has now been mutually agreed by the author and the publishers, Pergamon Press, that the need for this book has now passed. Pergamon have agreed that the material may be further used and, as a result, some of this may be found in the present publication in an updated form.

# CONTENTS

# 1

# Principles

ORAL medicine is generally understood as being the study and non-surgical treatment of the diseases affecting the oral tissues, especially the oral mucous membrane, but also other associated tissues and structures such as the salivary glands, bone, and the facial tissues. The boundaries of oral medicine are poorly defined; for instance, the investigation of facial pain and other neurological disturbances may be considered to be in the field of either oral medicine or of oral surgery. However, in the present book the discussion will be restricted to soft tissue lesions in and about the mouth, and to the effect of systemic disease on the oral tissues and related structures. It is the responsibility of the general practitioner to treat some of these conditions; others are often better treated in specialist clinics, but the general dental practitioner, to a very great extent, bears the responsibility for the recognition of oral disease at an early stage.

Perhaps the most important role of those working in the field of oral medicine is in the recognition of changes in the oral cavity resulting from generalized disease processes; many oral lesions which, in the past, were considered to be of entirely local origin are now known to be associated with systemic abnormalities. The most potent factor in the recent expansion of the scope of oral medicine has been the change of emphasis from the purely descriptive to the investigative. The modern concept of the subject implies a recognition of basic aetiological factors, of the histopathological changes occurring in the involved tissues, and of the significance of such matters as the general medical status of patients. As recently as 1955, Cooke pointed out in his study of leukoplakias and related lesions that previous workers had virtually ignored the significance of histopathology in their assessment, thus making it impossible to apply any but purely descriptive criteria to the conditions involved. The development of the discipline of oral medicine has depended largely on the adoption of an analytical approach based on the application of fundamental principles such as those mentioned above. It follows that the practice of oral medicine as a specialty depends largely on the availability of diagnostic facilities, often greater than those available to the general dental or medical practitioner, or even to some practitioners working in a hospital environment.

## Normal oral mucous membrane

In its basic structure the oral mucous membrane resembles other lining mucous membranes, for example, those of the vagina or the oesophagus, although within the mouth there is a wider range of epithelial structures than that seen in these other sites. These variations depend largely on differences in the degree of keratinization shown by the mucosae in different areas of the mouth. However, some of the reactions of the oral mucous membrane resemble those of the skin; this presumably is because of its position in the transition area between the gastro-intestinal tract and the skin. As a result of this, diseases both of mucous membranes and of the skin may produce lesions in the mouth. However, the oral mucosa characteristically behaves as a mucous membrane, its behaviour in disease processes perhaps most closely resembles that of the vaginal mucosa.

The oral mucous membrane consists both anatomically and functionally of two layers, one (the corium or lamina propria) essentially of mesodermal origin and one epithelial (Fig. 1.1). When considering variations of structure the behaviour of the corium must be taken into account even though the major changes may appear to be within the epithelial layer. In normal mucous membrane the integrity of the epithelium is maintained by the division of cells at or near the basal layer. As each cell divides one resulting cell remains effectively *in situ*, whilst one migrates towards the surface undergoing various structural modifications until it reaches the surface (Fig. 1.2). These modifications, which are dependent on the process of keratinization, vary according to the precise site of the mucosa involved and result in the production of a surface layer of cells which are either fully, partially, or non-keratinized and which are shed into the oral cavity at a rate dependent on the rate of mitosis at the basal layer. For each dividing cell one cell is lost from the surface, and thus, the integrity and dimensions of the epithelial layer are maintained. The rate of turnover of the surface cells has not been determined in man, but in animals it is known that it is high, the surface cells being replaced every 2 or 3 hours. This is considerably faster than in skin and must play a part in the defence mechanism of the oral

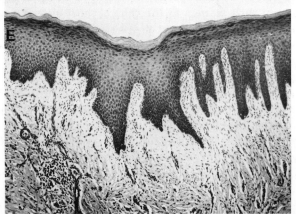

**Fig. 1.1.** Normal mucous membrane from the hard palate showing a surface layer of keratinized epithelium (E) lying over the corium (C).

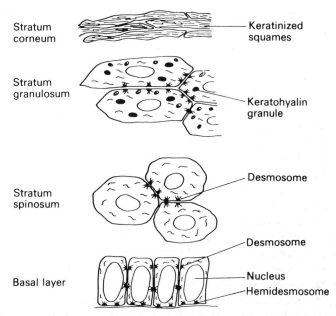

Stratum corneum — Keratinized squames

Stratum granulosum — Keratohyalin granule

Stratum spinosum — Desmosome

Basal layer — Desmosome / Nucleus / Hemidesmosome

**Fig. 1.2.** Diagram of a keratinizing squamous epithelium. Compare with Figs. 1.1, 1.3, and 1.4.

cavity against infecting organisms which are denied a stable site in which to proliferate.

The similar structuring of the epithelial layer of the skin has been shown to be maintained by a series of regulating mechanisms, chemically mediated, some of which are intrinsic to the epithelium and some concerned with mesodermal–epithelial relationships. Recent work has demonstrated the existence of 'chalones', chemical regulators of mitotic division which are produced within the epithelium and can be chemically separated. Other chemical mediators have been described which are produced within the mesodermal tissues and which exert a strong controlling influence on the structure of the overlying epithelium; it would seem that these mesodermally produced factors are active in maintaining the orderly arrangement of the epithelium from the basal layer to the surface. All recent work has strongly suggested that a similar series of complex and interacting factors operate in the oral mucosa, emphasizing the fact that any study of a mucosal lesion must include consideration not only of the epithelial tissues, but also of the underlying corium. This is so even when the lesions concerned are those often considered as being entirely epithelial as, for instance, leukoplakias and related lesions.

The epithelium of the oral mucosa shows wide variations in the extent of the keratinization process. In the fully keratinized situation the rather cubical cells formed by mitosis at or near the basal layer migrate towards the surface,

becoming more polyhedral and sharing intercellular attachments which have
given the name 'prickle cell layer' (or stratum spinosum) to this zone (Fig. 1.3).
In the light microscope these intercellular 'prickles' appear as single attachments
of the cell walls, but by electron microscopy these intercellular junctions
(referred to as desmosomes) are seen to be of much greater complexity (Fig. 1.4).
It is probable that the desmosomes act in a mechanical manner to give strength
to the epithelium; in several diseases marked by epithelial fragility the desmosome attachments are lost or impaired. It should perhaps be added that similar,
one-sided structures, hemidesmosomes, attach the plasma membrane of the
basal cells to the lamina densa of the basement membrane complex (Fig. 1.5). As
the cells of the stratum spinosum migrate to the surface they begin to flatten and

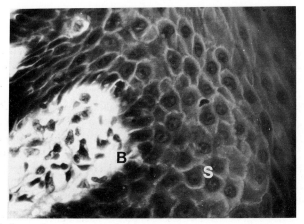

**Fig. 1.3.** Epithelium of oral mucosa showing basal layer (B) and the prickle cell layer (S)
(the stratum spinosum).

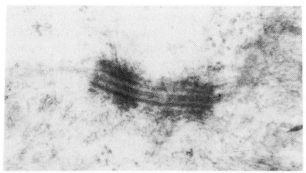

**Fig. 1.4.** Electron micrograph of a desmosome—the intercellular connection in the basal
and prickle cell layers of the epithelium.

Basal cell

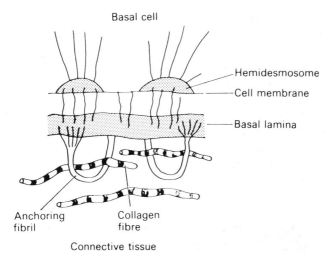

Hemidesmosome

Cell membrane

Basal lamina

Anchoring
fibril

Collagen
fibre

Connective tissue

**Fig. 1.5.** Diagram of the basal complex of the oral epithelium. The connections between the basal cells and the underlying connective tissue are via the hemidesmosomes.

granular structures (keratohyalin granules) appear within them. The origin and function of these granules are as yet undetermined, but it is known that they are closely involved with the process of keratinization. These granules give the characteristic appearance to the 'stratum granulosum' in keratinized epithelia. Finally, at or near the surface, the epithelial cells loose their detailed inner structure, the nuclei degenerate, the keratohyalin granules fragment and disappear and the insoluble protein complexes mentioned above fill the cell, now fully keratinized (Fig. 1.6). At this stage the desmosomes have effectively degenerated also and the flattened cells ('squames') are eventually lost into the oral cavity. As has been pointed out, each keratinized cell lost in this way must be matched by a dividing cell in the proliferating compartment of the epithelium in order for stability to be maintained.

This process applies only to fully keratinized epithelium—as seen, for instance, in the mucous membrane overlying the hard palate—and is usually referred to as orthokeratinization. In other areas (as in some parts of the buccal mucosa and the floor of the mouth) this process of keratinization does not take place, keratohyalin granules are not formed and nuclei and organelles (although somewhat effete) can be seen in the surface layers. In an intermediate form (parakeratotic epithelium) nuclei may still be seen in the surface layers and keratohyalin is sparse or absent, but some of the chemical changes of keratinization occur in the superficial cells (Fig. 1.7). Many workers have studied the varying processes of keratinization of the epithelium of the skin and oral mucous membranes, and varying classifications, often depending on histochemical staining reactions, have been proposed. However, for the purpose of understanding the clinical significance of these differences it is suggested that they should be

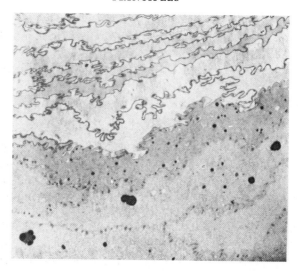

**Fig. 1.6.** Electron micrograph of granular cell layer of the oral epithelium (below) and the surface layer of keratin.

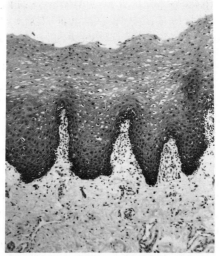

**Fig. 1.7.** Epithelium from the floor of the mouth showing parakeratosis. Nuclei are present in the surface layers.

regarded as being part of a spectrum ranging from complete non-keratinization at one extreme, through varying degrees of parakeratinization, to full orthokeratinization at the other.

The distribution of these differing epithelia in the normal oral mucosa has a close relationship with the function of the tissues at the site. In the normal situation, non-keratinized or parakeratinized epithelium is seen on the buccal mucosa, the floor of the mouth, and the ventral surface of the tongue, whilst orthokeratinized epithelium is seen on the hard palate and parts of the gingivae. The dorsal surface of the tongue is also orthokeratinized, but differs from the other oral mucosal surfaces in that there are a number of specialized structures present, predominantly the papillae. These latter (particularly the filliform papillae) are of considerable clinical significance in that their atrophy is often an early sign of mucosal abnormality.

Apart from the keratinocytes—the main cell component of the oral epithelium—there are other cells whose function and origin is still under intensive investigation. The two most important groups of cells of this kind are the melanocytes and the Langerhans cells. Both these types of cells appear in normally stained (H&E) sections of the oral epithelium as 'clear cells' in which the nucleus is surrounded by a clear zone of cytoplasm, but special stains (and electron microscopy) show considerable differences from the majority of the cells of the epithelium. These cells are dendritic, with cytoplasmic prolongations extending between the cells of the basal and suprabasal areas of the keratinocytes, but with no attachments to them. Other classes of dendritic cells have been demonstrated, with different characteristics, but the function of these is not yet known.

The melanocytes appear in, or very close to, the basal layer and on electron microscopy show granular structures (melanosomes) which are the precursors of melanin: the black pigment which modifies the colour of both skin and mucous membranes. The melanotic pigmentation of oral mucosa shows great racial variation, in parallel with that of the skin. However, this does not depend on variation of the numbers of melanocytes but on the number and activity of the melanosomes within them. The epithelia of all races contain approximately the same number of melanocytes; it is their activity which varies. It is known that hormonal influences are important in the stimulation of melanocyte activity, although the precise mechanisms remain obscure. In particular, the function of the so-called melanin-stimulating hormone (MSH) has, as yet, not been fully elucidated. In some circumstances the melanocytes may be stimulated to produce excess melanin by a wide range of non-hormonal stimulae. This will be discussed in later chapters.

The Langerhans cells were for long something of a mystery. There are a substantial number of these cells present near the basal complex of the oral epithelium, with dendritic processes extending between the keratinocytes and with recognizable ultrastructural features. In the previous edition of this book the Langerhans cells were described, in the light of the then available research reports, as cells which in some way take part in the regulation of the keratinization process. This view has changed and it is now suggested that these cells have an immunological function, acting as peripheral scavenging cells of the immune system not unlike macrophages. It would seem that at least one function of these cells is to accept antigens and stimulate the activation of T lymphocytes against them. Recent work has suggested that there are variations from the normal both in the number and the immunological reactivity of the Langerhans cells in some lesions of the oral mucosa; for instance, in oral lichen planus and in candidal leukoplakia.

The role of iron metabolism in the maintenance of the structure of the oral mucosa has been the subject of much investigation. It is certainly the case that iron deficiency, even when relatively early in clinical terms, can result in generalized oral epithelial atrophy and loss of the papillary pattern of the lingual mucosa. It seems that other deficiencies which might affect iron metabolism and erythrocyte production, such as folate and B12 deficiencies, may also contribute to this destabilization of the oral epithelium. This will be discussed at greater length in Chapter 9.

Lying deep to the corium of the oral mucous membrane is the submucosal layer, separated from the corium by a gradual zone of transition rather than by a clear boundary. The submucosal tissue components are widely variable: blood vessels, fat, and fibrous tissue being present in differing proportions according to the precise site. In the corium and submucosa lie the minor glands and sebaceous glands of the oral cavity; again, these are widely variable in distribution, the mucous glands being most frequent in the mucosa of the lips and posterior palate whilst the sebaceous glands are mostly concentrated in the buccal mucosa. Within the corium and submucous tissues are scattered cells of the leukocyte series in varying proportions and concentrations. During disease processes these may alter radically, both in number and in type, depending on the basic nature of the pathological process involved. There is, in fact, evidence that in certain diseases of mucosa and skin in which the apparent abnormalities are epithelial such alterations in the subepithelial leukocyte population may represent the initial pathological change (lichen planus is a good example of this).

Lying between the epithelium and corium of the oral mucous membrane is a dividing structure, the basement membrane. When viewed in the light microscope this appears as a relatively substantial layer, but the electron microscope has shown that this appearance is deceptive. The 'basement membrane' of the light microscopist is, in fact, a zone of biochemical activity as can be demonstrated by a number of special stains. On ultrastructural study it is seen that the components of the basal zone are much finer than suggested by light microscopy and that, rather than a single membrane, at least two zones are visible (the zona lucida and the lamina densa). In this area fibres attach the lamina densa to the underlying tissue and, probably, to the hemidesmosomes of the basal cells of the epithelium (Fig. 1.5). The origin of the basal complex is not yet decided. Until recently, it was thought to be of mesodermal origin, but current opinion is that the lamina densa and the lamina lucida are derived from the epithelium. As yet, the behaviour of the basal complex in varying pathological states is not fully assessed but there is no doubt that many oral mucosal changes are associated with abnormalities in this area. In particular, the advance of immunological techniques at the ultrastructural level is leading to detailed studies of the behaviour of the basal complex in disease processes. For instance, the existance of pemphigoid antigen sites has been described at which circulating antibodies, found in patients suffering from pemphigoid, may react. This results in weakening of the epithelial–connective tissue junction and the production of bullous lesions (see Chapter 8).

Although the oral mucous membrane has several functions, sensory and secretory among them, its main purpose is probably that of acting as a barrier. Recent work implies that there may be two sites at which this barrier function may occur. One of these is at the basal complex and the second is in the intercellular substances of the middle layers of the epithelium. At both these sites it appears that some molecules are selectively blocked from passing inwards into the tissues. In considering this protective function it is also necessary to discuss other factors, in particular the role of saliva. The oral mucosa is constantly bathed by saliva which not only maintains the physiological environment necessary for the maintenance of epithelial integrity but also includes a number of protective, antibacterial components. A number of these have been described, but perhaps the most important are the secretory immunoglobulins, predominantly of the IgA class, which are found in saliva and which attach to sites on the epithelial surface (see Chapter 2). It seems possible also that salivary mucosubstances form a physical coating which remains intact over the oral epithelium in the healthy individual and which may also exert a protective action.

In spite of this barrier function of the oral mucosa there is a degree of permeability which, apart from its theoretical and scientific interest, is also of clinical significance. During local therapy with mouth washes and similar preparations, drugs may be transported across the oral mucosa and may exert effects similar to those resulting from systemic therapy. This factor must evidently be considered if (for example) therapy with high concentration steroid mouthwashes is to be carried out. The permeability of the oral mucosa to drugs is most commonly utilized in the treatment of angina by glyceryl trinitrate and similar substances. In these circumstances the very rapid absorption of the drug is of obvious advantage. It is thought that, apart from the mechanism of simple diffusion, active transmission of some substances by cell-mediated mechanisms may occur. It has also been suggested that surface active agents such as are commonly used in toothpastes and mouthwashes may influence the permeability of the oral mucosa to other substances. There is currently a great deal of research activity in this field.

Although the full significance of the role of saliva in maintaining the health of the oral mucosa is, as yet, not fully understood there can be no doubt that a free salivary flow is an essential part of the oral environment. If the flow is diminished, either by degenerative changes in the salivary glands or by the action of drugs, soreness and atrophic changes in some areas of the oral mucosa rapidly follow. The tongue is perhaps most markedly affected in this way. In some conditions (for example, Sjögren's syndrome—Chapter 5) it is difficult to distinguish between primary mucosal changes and those secondary to diminished salivary flow, but on a clinical basis it is reasonable to accept that atrophic changes in the oral epithelium are regularly associated with dryness of the mouth.

A further component to be considered as part of the normal healthy oral environment is the microbial flora of the mouth. A wide range of organisms may be present in the oral cavity, living in a commensal relationship with the host. When this relationship is upset by a change in the local or generalized conditions then the commensal organisms may become pathogenic. A number of the more common oral infective conditions represent such a variation in the host resistance to organisms normally present in a commensal state. Acute ulcerative gingivitis is an example of this, although the precise change in the host leading to clinical infection is not easy to identify. It need hardly be said that some other infections of the oral cavity (for instance, syphilis) result from straightforward primary inoculation by external pathogenic organisms to which the patient has little or no resistance. A more complete discussion of the aetiological factors in some common oral infections is given in Chapter 3.

## Age changes

It is generally accepted that changes occur in the oral mucosa of healthy individuals with increasing age. Reductions in overall epithelial thickness, flexibility of the collagen fibres, innervation, blood supply, and permeability of the mucosa have been described. It has generally been assumed that changes of this kind occur with comparative suddenness at a late age, but recent evidence implies that, at least in the case of the epithelium of the tongue, a continuous trend towards atrophy occurs throughout adult life. The clinical significance of this observation is that sudden changes in the structure of the oral mucosa (for example, depapillation of the tongue) should never be considered as being due to age alone without full investigation and the elimination of other factors: haematological, nutritional, and so on (see Chapter 5).

A further age change which may affect the function of the oral mucosa is in the salivary glands. It has been shown that gradually increasing degrees of atrophy and fibrosis affect the secretory units of both the submandibular and the labial salivary glands throughout life, even in the absence of disease processes which might be associated with such change (Chapter 5). As considered above, the resulting decrease in salivary flow from these glands may well affect the function of the mucosa as a whole. However, this is an area of current doubt since it has also been shown that, in general, parotid salivary flow remains unaffected by age change alone. The clinical aspects of this problem are further considered in Chapter 5.

## Abnormal oral mucous membrane

Many oral lesions represent the end result of breakdown or abnormality of the normal structuring of the epithelium. Variation in the rate of keratin formation, disproportion between the different layers of the cells, breakdown of the normal intercellular bonds of the prickle cells, splitting of the epithelium from the connective tissue, and many other similar abnormalities may occur in different diseases. For instance, in a number of mucosal abnormalities hyperkeratosis occurs (Fig. 1.8). This may arise as a result of abnormal irritation of the mucosa or apparently spontaneously in some conditions. In other lesions, atrophy of the epithelium may occur. This represents a thinning of the normal epithelial layer, perhaps to only a few layers of cells, often accompanied by incomplete keratinization. Such epithelium is easily lost following a minor degree of trauma and so atrophic lesions of the mucosa readily become ulcerated (Fig. 1.9). Many of the so-called erosive lesions are of this type. It should be remembered that ulceration is in itself a quite unspecific process and implies only the loss of epithelium from the mucosal surface followed by inflammatory changes in exposed connective tissue. Bullae or blisters of the mucosa may occur in one of

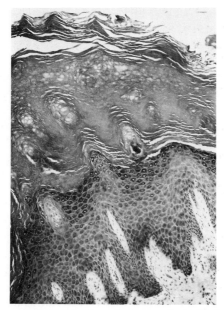

**Fig. 1.8.** Hyperkeratosis—the production of an excessively thick layer of orthokeratin.

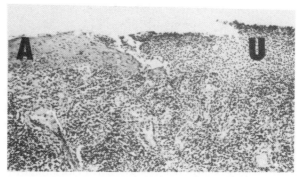

**Fig. 1.9.** Atrophic epithelium (A) which has been lost over part of the area with resulting ulcer formation (U).

two ways, either by degeneration of the cells and of the intercellular links in the prickle cell layer of the epithelium (Fig. 1.10) or by separation of the whole of the epithelium from the underlying corium (Fig. 1.11). In the former case, the bulla is entirely within the epithelium and may include within its contents the rather rounded cells which result from the degeneration of the prickles. These cells may be examined in a diagnostic smear and the process (known as acantholysis) recognized (Fig. 1.12). On the other hand, the bullae formed by the lifting off of the epithelium from the corium will contain only inflammatory cells and no acantholytic epithelial cells. This difference is important in the differential diagnosis of bullous lesions in the mouth. The immunological features leading to

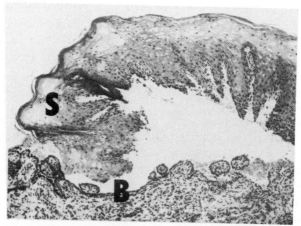

**Fig. 1.10.** A bulla (or blister) formed as a result of loss of intercellular cohesion. The bulk of the epithelium (S) has separated off to form a bulla, leaving the basal cells (B) to form the base of the lesion. This is an intra-epithelial bulla.

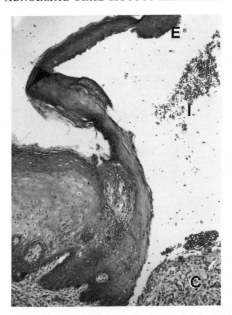

Fig. 1.11. A bulla formed by the separation of the entire epithelium, including the basal cells (E) from the underlying corium. The bulla fluid contains inflammatory cells and red blood cells (I). This is a subepithelial bulla.

the production of lesions of this kind are discussed in Chapter 8. It must be reiterated that changes seen in the oral epithelium are not confined to the epithelium alone. Frequently, there are also changes in the supporting tissues and, in some cases, the visible epithelial changes may be secondary to changes in the underlying corium which affect the nutrition and metabolism of the epithelium. The greatest practical significance of this fact is, perhaps, the necessity when taking a biopsy of lesions of the oral mucosa, to include a representative thickness of corium in the tissue removed for microscopic examination. In many cases, a biopsy consisting largely of epithelium alone is virtually useless for diagnosis.

The integrity of the oral mucosa is maintained by a complex of interacting factors superimposed on the localized stabilizing mechanism discussed above. The general hormonal status of the patient, and a number of nutritional and metabolic factors are involved in maintaining the cell metabolism and the ordered structure of the mucous membranes. If any single factor is disturbed then sequential changes occur and clinically significant abnormalities of the oral mucosa may follow. It is often difficult to decide which of the various possible factors are involved in initiating these changes—these may evidently occur either as a primary manifestation of localized mucosal abnormality or as a secondary effect of generalized disease processes. It is the function of those

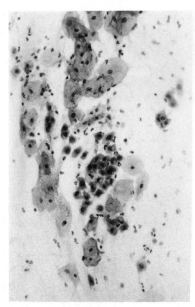

Fig. 1.12. A smear made from the bulla fluid in a case of pemphigus. Although many inflammatory cells are present the rounded acantholytic epithelial cells of pemphigus can be seen.

working in the oral medicine clinic to assess the possible aetiological factors associated with mucosal lesions of this kind and to ensure appropriate investigations and (if needs be) treatment.

The reactions of the oral mucosa are not exclusively those of a mucous membrane; as has been pointed out, a number of diseases of the skin also find expression in oral lesions. This is not entirely surprising on anatomical grounds since the larger part of the oral mucosa is derived from an embryonic invagination which carries inwards some of the precursor epithelial cells from which both facial skin and oral mucosa are developed. As might be expected, the lesions of oral mucosa and skin which occur in these mucosal–cutaneous diseases are often superficially different although the basic histological changes seen in the tissues are similar. Such differences are seen in the primary lesions and, presumably, depend on the differences between the structure of the mouth and of the skin. Quite often secondary changes also occur in oral lesions. The continually wet environment of the mouth, in combination with repeated mild trauma of the tissues by teeth and foodstuffs, and the presence of a wide range of microbial flora further modifies the nature of the lesions produced in a number of diseases; for instance, should the epithelium be thinned by atrophy or weakened by the formation of blisters, it is likely to be lost and the initial lesion be replaced by an ulcer. For reasons such as these, oral lesions, particularly at an advanced stage, may show less characteristic features than the equivalent skin lesions of the same disease. Clinical diagnosis in such circumstances may be quite difficult since only areas of ulceration of a relatively non-specific nature may be present rather than fully developed specific lesions. For the diagnosis of an oral mucosal disease by histological criteria, it is often necessary to await the appearance of new lesions and to examine these at an early stage before the secondary changes occur.

## Histological changes

It may be helpful to recall some of the terms used to describe changes seen on histological study of the oral mucosa.

**Hyperkeratosis**—an increase in the thickness of the keratin layer of the epithelium, or the presence of such a layer in a site where none would normally be expected (Fig. 1.8). Hyperorthokeratosis is the term used to specify a thickened, completely keratinized layer, whilst in hyperparakeratosis there is incomplete ketatinization with nuclei remaining in the surface cells.

**Acanthosis**—an increase in thickness of the prickle cell layer of the epithelium. This may or may not be accompanied by hyperkeratosis.

**Atrophy**—a decrease in the thickness of the epithelium, often associated with incomplete keratinization (Fig. 1.9).

**Oedema**—the collection of fluid in or between the prickle cells, intra- or intercellular, the two forms often occurring simultaneously. Oedema may also occur between the epithelium and the corium in the region of the basal complex.

**Acantholysis**—loss of the intercellular attachments in the prickle cell layer leading to separation of the cells. When associated with intercellular oedema this leads to the production of intra-epithelial bullae (or blisters; Fig. 1.10).

**Atypia**—this is a compendium term used to describe variations in the maturation of the epithelial cells which may be associated with malignancy or premalignant potential. Such features as abnormal mitoses and lack of normal structuring of the epithelium are taken into account in the assessment of atypia.

# Immunity

The basic concept of immunity has been known for several hundred years, since it was first realized that a single attack of a disease such as smallpox or measles is sufficient to provide lasting protection against a recurrence of the same disease. In the early nineteenth century the recognition of bacteria, and of the process of infection and resistance led to the idea of immunity as an essentially antimicrobial process—a view which lasted until relatively recently. It has now been accepted, however, that the immune response is a process in which a wide range of foreign or abnormal material, not necessarily microbial in origin, may be recognized and immobilized so as to eliminate the risk of damage to the host. Put rather differently, the immune response is a reaction by which the host animal is able to recognize and reject foreign or abnormal material ('non-self') whilst accepting its own normal materials ('self'). Among the fundamental concepts laid down by the early workers in immunology was that of 'horror autotoxicus'. the idea that, since 'self' is acceptable to the immune defence system, antibodies may never be produced to act against the host's own tissues. It is now quite clear, however, that in some circumstances, the mechanisms of the immune response may, in fact, be directed against the host in this way. This lack of recognition of 'self' is the fundamental process occurring in a number of diseases (the autoimmune diseases), the process being known as autoimmunity.

Among the human diseases which may be explained in terms of abnormalities or deficiencies of the immune mechanism are a number which involve the oral tissues and it is important for those with an interest in oral medicine to have an appreciation of the processes involved.

## Antigens and the immune response

An antigen is a substance which, when introduced into a host animal, stimulates a protective immune response directed towards the elimination or neutralization of the antigen. A large number of compounds or parts of compounds may act as antigens; most are proteins (or parts of protein molecules), but some carbohydrates and lipids may act in this way. The property which determines the antigenic nature of a substance is its foreign nature in relation to the host; a substance which acts antigenically in one species may not do so in another species. Similarly, substances normally present in one species of animal may act as antigens in other species.

When an antigen is encountered by the host for the first time, the response is relatively minor—the primary response. Following this, however, re-exposure to the same antigen may result in a much more vigorous reaction—the secondary response. The reason for this is that many of the lymphocytes which are involved in the primary response remain as 'memory cells' and maintain their responsive state to the specific antigen. It is known that this 'memory' is a function of the lymphocytes since in suitable circumstances immunological activity may be transferred between animals by the transfusion of lymphocytes.

When antigens are introduced into an animal two types of immune responses may occur. One of these (the humoral response) consists of the production of antibodies which are able to react with and neutralize the effect of the antigen. Antibody activity is associated with the globulin fraction of the plasma proteins, from which a number of distinct groups, known as immunoglobulins (abbreviated Ig), may be separated. These groups are distinguished by a suffixed letter and the main immunoglobulins are known ad IgG, IgM, IgA, IgD, and IgE. Apart from the chemical distinction of these immunoglobulins (depending on such factors as molecular weight), each type has well defined immunological characteristics and takes part in different aspects of the immune response. Thus, IgG is the most important immunoglobulin in the neutralization of bacterial toxins whilst IgA is selectively secreted in saliva, tears, and mucus, and plays a major part in the defence of skin and mucous membranes against bacterial attack. Characteristic variations of the serum immunoglobulins may take place in different groups of diseases and, although the changes are not specific, it is often possible to gain some insight into the origin of a disease process by the study of the immunoglobulins circulating in the affected patient.

The synthesis of antibodies is under the control of a section of the lymphocyte population known as the B lymphocytes. A lymphoid structure, the bursa of Fabricius, has been recognized in birds as being involved in the development of B lymphocytes and, although the equivalent organ has not been identified in man, the term B (bursa-dependent) lymphocyte is universally used to describe this group of cells. It would seem that the bursa equivalent in man is probably represented by a diffuse distribution of lymphoid tissue throughout the gut, including such structures as Peyer's patches in the small intestine and, possibly, the tonsils. When stimulated by contact with an antigen, the B lymphocytes proliferate in the peripheral lymph nodes and become transformed into plasma cells which actively synthesize and secrete the antibodies.

When antibodies react with antigens a third factor often affects the result. This is the presence of complement, a complex of at least nine plasma proteins (not immunoglobulins) which react in a sequential manner when activated by an immune complex (that is, the combination of an antibody and an antigen). The first complement component is activated by the complex and each molecule so activated in its turn activates several molecules of the next component. This escalation is maintained through the whole series of components of the complement—the so-called complement cascade. At each stage of the cascade, enzymes and other biologically active substances are released. These produce varying effects, but all act to reinforce the immune response to the antigen. Thus, some of the released factors act chemotactically by attracting polymorphs to the area and so reinforce the inflammatory response, whilst others act directly on the walls of foreign cells or bacteria, making them permeable and susceptible to damage. The actual mechanism of the complement cascade is known to be highly complex and to involve several alternative pathways of action, but for practical purposes, there is little to be lost by regarding the complement cascade as a single process which reinforces the mechanism of the other branches of the immune response.

The second main type of immune response, the cell-mediated response, depends on direct cellular activity between lymphocytes and antigen. The lymphocytes involved in cell-mediated immunity are known as T lymphocytes since they are dependent on the presence of the thymus in the same way that the B lymphocytes are dependent on the bursa of Fabricius or its equivalent. When T cells become sensitized by exposure to an antigen they will react, on subsequent exposure to the same antigen, by transforming themselves to an active form, lymphoblasts. This transformation is accompanied by a great deal of cellular proliferation and activity and by the release of a number of active soluble substances (lymphokines). The active lymphoblasts react directly with the stimulating antigen whilst the lymphokines act in a number of ways to enhance this effect. Both B and T lymphocytes are initially derived from the stem cells of the bone marrow and it is not possible to differentiate between them on histological grounds. They may, however, be identified by techniques which depend on their immunological characteristics and by transmission and scanning electron microscopy. Although the humoral and cell-mediated immune responses are in many ways distinct there is considerable co-operation between them and it has been established that there are many situations in which it is necessary for both systems to function in order to produce a satisfactory response.

Not surprisingly, the simple division of the lymphocyte population into two major divisions has been complicated by the recognition of subdivisions and also of 'null cells' which have a range of functions, including 'killer' activity, some in the absence and some only in the presence of antibodies. T cell subgroups include cytotoxic cells, which are particularly active against cells containing viruses, and helper cells, which co-operate with B lymphocytes in the overall immune response. Many other subgroups, recognized both by relatively major or by minor markers (such as surface antigenic sites), have been described and their probable functions outlined. Just as in the case of T lymphocytes, B lymphocytes have been divided into groups, although not in such a complex a manner as have the T lymphocytes. A number of the subgroups of B cells have been described as a result of complex laboratory tests rather than from evident behavioural differences, but perhaps the most clinically significant differentiation is between T cell dependant and T cell independent lines.

There are two quite separate areas in the oral cavity in terms of immunological activity. The first, and by far the larger, is the area in which saliva is the dominant fluid element. Clearly, this includes most of the oral cavity and its mucosa. However, a second area, although much less in extent, is very important in terms of its impact on dental and periodontal disease. This is the area of the gingival crevice in which the fluid component is largely derived from exudates through the crevicular epithelium. As has been pointed out above, the main immunologically active component of the saliva is IgA. This is present in the saliva in a much higher ratio to that of other immunoglobulins than in serum. It is currently thought that this IgA is produced by B lymphocytes within the salivary glands and that it is not the result of concentration of serum IgA by the glands. The main cellular component of saliva (apart from such inactive constituents as desquamated epithelial cells) is of polymorphoneuclear leukocytes, derived from the crevicular fluid, the majority of which are effete.

The situation is quite different in the crevicular area where there are significant concentrations of IgG, IgM, and IgA as well as components of the complement series, all derived from the crevicular fluid. Cellular components include an active polymorph population, macrophages and lymphocytes of both B and T types. Clearly, the environment in the gingival crevice, in terms of immunological activity, is quite different from that in the rest of the oral cavity. The complex immunological changes which may take place in this area, involving both enhancement and suppression of differing processes, are under constant review as part of many research programmes into the aetiology of both chronic periodontal disease and caries.

A few individuals are born with gross immunodeficiencies. These are largely genetically determined and the affected individuals tend to die young, often as a result of recurrent infections. These major conditions are fortunately rare, but many minor deficiencies of the immune mechanism are known and a number of these produce effects in and around the mouth. Acute ulcerative gingivitis, candidiasis, and recurrent herpes are examples of conditions which occur in or around the oral cavity and which may be related to varying degrees of anomaly in the immune system.

Recent work of great interest concerns the complex inter-relationships between the immune responses to neoplastic and infective antigens in patients with neoplasms. The relationship between infection and neoplasia has not been clearly established in man, although numerous virus-induced animal neoplasms are known. An example of human neoplasm suspected of a close association with viral infection is the so-called African lymphoma or Burkitt tumour. Patients with the lymphoma have been shown to have raised levels of antibodies to the Epstein–Barr virus, a type of herpes virus which can be demonstrated in the tumour cells. It has also been shown that children living in the areas of the world in which this neoplasm occurs and who have high serum levels of antibody to the virus run a much higher risk of eventually developing the lymphoma. The remarkable feature of this virus–neoplasm association lies in the fact that the virus has been shown beyond reasonable doubt to be that which, in other geographical locations, is responsible for the very different disease, infective mononucleosis. It is suggested that, whilst infective mononucleosis is the result of infection of B lymphocytes, the Burkitt lymphoma represents a proliferative response to T lymphocyte infection. This, however, does not provide an explanation of the restricted geographical distribution of this neoplasm. The suggestion has been made that the induction of the neoplasm is dependant on the lowering of the defence mechanism due to a second infection—malaria. If so, this provides a parallel situation to that of cancrum oris in which it is suggested that an explosive extension of acute ulcerative gingivitis in susceptible patients may occur following a viral infection (usually measles) which causes a transient suppression of the immune system (see Chapter 3). Recent work on oral leukoplakias seems to point in a similar way to a deficiency in the response of some of the patients to infection by Candida albicans and herpes, which is associated with a minor degree of antigenic activity in the leukoplakias involved. There is, of course, a known association between candida infection and leukoplakia, and a tentatively established one between the herpes virus and the

induction of carcinoma. It is, therefore, clear that the accumulation of evidence
to link the three factors would be of great importance.

However, quite clearly, the current predominant interest in the field of the
effects of virus—induced immune suppression within the oral cavity lies in the
HIV virus and the oral manifestations of AIDS. This subject will be considered in
Chapter 3.

## Hypersensitivity

In contrast to the diseases caused by immune deficiencies there are others which
depend on an over-active or exaggerated response of some aspect of the immune
system. The conditions known as hypersensitivity reactions are of this type, and
they depend on an enhanced response of either the humoral or the cell-mediated
mechanisms consequent upon a second contact with an antigen to which the
host has previously been sensitized. These reactions include some in which a
severe toxic effect may be produced in the host.

The significance of these hypersensitivity reactions in oral medicine is two-
fold. In the first instance, many of the conditions resulting from the reactions
lead to secondary oral symptoms or affect the oral tissues as part of a more
diffuse and generalized process. In some circumstances, as, for instance, in
erythema multiforme, the oral symptoms, although part of a generalized
reaction, may prove clinically to be the most significant. The second important
group of hypersensitivity reactions affecting the oral tissues is that which is
involved in the production of chronic periodontal disease. These processes have
recently been carefully examined and it would appear that hypersensitivity
reactions may play a part in the initiation and continuance of chronic gingivitis
and periodontitis. The antigenic stimulus is thought to be in this case provided
by the organisms present in the plaque, the resulting hypersensitivity reactions
being responsible for tissue damage or destruction in the periodontal tissues.

## Autoimmunity

In an autoimmune reaction the immune response is directed against the host's
own tissues which, for some reason, have become antigenically active. Both the
cell-mediated and the humoral responses may be involved in the process. It is
not known how the facility to recognize 'self' is destroyed, but a number of
theories have been proposed. In the context of oral disease it has been suggested
that tolerance to 'self' antigens of the oral epithelium may break down in the
face of repeated stimulation of the local immune system following minor tissue
damage or infection. Such provocation might occur, for example, during tooth
eruption or in minor periodontal disease.

A number of diseases of accepted autoimmune origin affect the oral tissues
secondarily; thus, in pernicious anaemia (Chapter 9), although the primary
autoimmune process directly affects the gastric parietal cells, the haematological
changes induced by the resulting inability to absorb vitamin B12 may cause

marked abnormalities in the oral mucosa. These are the result of instability of the epithelium induced by the deficiency. Apart from these and similar secondary effects, however, a number of diseases of established or suspected autoimmune origin affect the oral tissues directly, producing oral lesions as a primary symptom. Of these, the most common is probably recurrent oral ulceration, the immunological aspects of which have been much investigated and will be more fully discussed in Chapter 3. Other examples of oral lesions in autoimmune disease will be described in the chapter dealing with the oral manifestations of skin diseases.

## The oral mucosa in generalized disease

Oral lesions may occur in a wide variety of generalized diseases. This fact is important, not only because of the need to treat the often painful oral lesions, but also in view of their significance in providing a diagnostic indicator. The mouth is readily available for inspection (and for biopsy) and oral lesions may appear early in some diseases. Thus, a number of important conditions may be first diagnosed following the proper evaluation of oral lesions.

The relationship between generalized and oral disease is a very complex one, but it may be helpful to identify four types of such inter-relationships.

1. In one group of conditions the oral lesions are similar in aetiology and histology to those found elsewhere in the body, being modified only by the oral environment. Many of the oral lesions of skin diseases fall into this group as well as those in some gastro-intestinal conditions. These are discussed in Chapter 8.

2. A second group of oral lesions result from changes in the metabolism of the tissue under the influence of abnormalities of nutrition, endocrine, and other factors. These abnormalities themselves are the result of some distant pathological process. The oral lesions associated with malabsorption fall into this group.

3. In this group both oral and more generalized lesions result from a systemic (and different) abnormality. Sjögren's syndrome, a manifestation of generalized autoimmune disease, is an example.

4. Interest is currently being shown in the occurrence of certain diseases in patients with specific genetic markers. The genetic system involved is the major histocompatibility system which determines the acceptance or rejection of transplanted non-self tissues. The genetic determinants are referred to as the HLA system (histocompatibility locus antigen system) which can be determined in any individual by the reaction of specific antibodies with antigenic sites on the lymphocytes.

The clearest association of this kind is between the HLA B27 locus and a group of arthropathies, in particular ankylosing spondylitis. However, a number of diseases affecting the oral mucosa including lichen planus, recurrent oral ulceration, and Beçhet's syndrome have been suggested as having specific HLA associations in common with other diseases. These associations are not yet proved, and, in fact, provide material for considerable scientific controversy. It is evident, however, that such a mechanism of association is of great interest in the context of oral disease in relationship to generalized disease.

It need hardly be said that this tentative classification of the oral–general disease relationships is oversimplified. Quite different classifications could easily be suggested. It is also evident that more than one form of association might occur in a single case. For instance, a patient with a disease of the lower gastro-intestinal tract might well produce primary lesions of the disease on the oral mucosa, as well as secondary mucosal change consequent on malabsorption.

# 2

# Investigations: therapy

## Methods of investigation

The basis of the investigation is a careful and detailed clinical examination. In the type of patient under consideration, an accurate case history is particularly important. It is also of great importance to obtain details of the medical history of the patient and of any current or recent drug therapy; very often the patient's general medical practitioner may be asked to fill in details. In the hospital environment, the request for and careful reading through of the patient's general hospital case sheet can be, by far, the most productive method of assessment in complicated cases. When dealing with the past medical history it is necessary to use direct questioning on some points. As an example, soft tissue lesions of the mouth may be associated with skin rashes, or eye or genital lesions. The connection with mouth lesions may seem quite tenuous to the patient who may very well fail to volunteer information on these points unless directly asked.

When examining the mouth, the whole of the oral mucosa must be carefully examined. All removable appliances should be taken out. The lips and cheeks must be gently retracted to display the full extent of the sulci and the tongue gently held with the aid of a gauze napkin, and extended forward and to each side. Care must be taken that the whole of the floor of the mouth and under-surface of the tongue is seen. The posterior part of the tongue, tonsillar fauces, soft palate, and part of the pharynx are exposed by gentle pressure on the tongue and helped by phonation of 'Ah' by the patient. This examination of the oral mucosa must be combined with a careful assessment of the other dental structures, facial skeleton, and soft tissues of the neck. Radiographs must be taken if indicated.

The quality of available light is particularly important in the examination of lesions of the oral mucosa. In general, the nearest approach to natural light is the most suitable. Many otherwise perfectly satisfactory dental lighting units are almost useless for the examination of lesions of the mucosa. In particular, those with a pink tinge make recognition of faint white lesions of the mucosa very difficult.

It is quite evident that a large number of diagnostic tests and procedures must be available to those working in the field of oral medicine. The advantages of close association with specialized departments, such as those of immunology, haematology, gastro-enterology, and so on are evident, and virtually all centres in which oral medicine clinics are successfully conducted enjoy such associations. Although the range of investigations carried out in the oral medicine clinic itself may be wide, it is evident that, in many instances, it is proper to refer the patient to a colleague in some other speciality for subsequent investigation following initial diagnosis in the oral medicine clinic.

Special investigations will be discussed subsequently, in the appropriate chapters and in relation to the subject of the investigation. However, the most common laboratory investigation carried out on patients in the oral medicine clinic is a screening procedure for possible haematological abnormalities. In view of its widespread application this merits preliminary discussion.

## Blood examination

Blood examination may be carried out either as a screening test to help in the diagnosis of some unidentified condition or to confirm a diagnosis by the application of specific tests. As a screening procedure for possible haematological abnormality, a haemoglobin estimation, white cell count, and blood film examination were formerly considered to be sufficient. However, it has been recently recognized that oral signs and symptoms may accompany relatively minor changes in the blood and that oral lesions may occur early in patients with haematological abnormalities, well before these are shown up by the simple examination of peripheral blood. It has also been demonstrated that in patients of this kind, mostly presenting with stomatitis or recurrent oral ulceration, significant lowering of serum or red cell folate levels or (less frequently) lowered serum B12 levels may occur in the absence of any detectable change in the peripheral erythrocytes. It therefore follows that the simple full blood count is an insufficient procedure for the initial investigation of such patients.

When a screening procedure is decided upon, a reasonable scheme of investigation is as follows.

(1) Full blood count, haemoglobin estimation, and film examination. From this, evident anaemias are demonstrated by variations in red cell morphology and lowered haemoglobin values. Abnormalities of the white cells are also shown.

(2) Estimations of serum iron, total iron binding capacity, and saturation. These provide an important test for latent iron deficiency. It has been suggested that serum ferritin estimations might be of significance in the evaluation of recurrent oral ulceration. However, as yet this test is not generally available and its full significance in terms of oral disease is not yet established. It is, however, an important test in the differentiation between patients with primary and secondary iron deficiencies.

(3) Serum B12 and folate and red cell folate estimations. These are valuable indicators of malabsorption and, hence, of gastrointestinal diseases of many kinds. The red cell folate level is a long-term indicator of folate deficiency,

whilst the serum folate levels are more labile and indicate the current status. It has been shown that these may show independent clinically significant variation in patients presenting with oral signs and symptoms. The older attitude that the red cell folate level should be estimated only in the presence of a lowered serum folate level is now known to be incorrect.

(4) As an additional test an erythrocyte sedimentation rate measurement is useful as a non-specific guide to underlying pathological processes such as chronic inflammatory conditions or neoplasia.

It is evident that economic factors alone preclude the full haematological screening of all patients attending with minor degrees of stomatitis or ulceration, but it is suggested that there are certain groups which warrant this action.

(1) Patients with persistent and troublesome oral ulceration.

(2) Patients with oral lesions with an atypical history or unusually resistant to treatment.

(3) Patients complaining of a sore tongue, a generally sore mouth, or abnormal taste sensation, even though no mucosal changes can be seen.

(4) All patients with candidiasis.

(5) Patients showing abnormalities following a simple full blood count or with a history of anaemia.

Since these groups constitute a large proportion of patients attending the oral medicine clinic, a full haematology screen is almost a routine procedure in this environment.

It is perhaps worthwhile considering the difference between the two estimates of folate status which are essential to give a comprehensive view of the situation. The red cell folate level is a relatively stable index of folate status, its levels varying relatively slowly in relation to the formation and destruction of erythrocytes in their normal cycle. Thus, the red cell level indicates the status of the patient at the time that the current population of red cells was formed. Serum folate levels, on the other hand, reflect the immediate situation and may vary on a day-to-day basis. A single folate-rich meal or a single dose of folic acid may be sufficient to transiently raise the serum folate levels from below normal to well into the normal range. It is generally assumed that patients with folate deficiency of any kind will be identifiable by an increase in mean cell volume (i.e. macrocytosis). It has, however, recently been shown that, at least in the patients attending the oral medicine clinic and, hence, predominantly complaining of sore mouths, the mean cell volume is not a good indicator of folate deficiency. Indeed, there may be normal erythrocyte morphology in the face of a clinically significant folate deficiency. This will be further discussed in Chapter 9.

In Table 2.1 the meanings of some of the terms used to describe variations from the normal in the size and shape of erythrocytes are given, together with an indication of some of the conditions in which these forms may occur.

**Table 2.1.** Variation in size and shape of erythrocytes

| Description | Erythrocyte characteristics | Seen in |
|---|---|---|
| Hypochromic | Pale staining | Iron deficiency anaemia |
| Hyperchromic | Dense staining | Pernicious anaemia |
| Microcytic | Small | Iron deficiency anaemia |
| Macrocytic | Large (10–12 μ) | Pernicious anaemia |
| Megalocytic | Very large (12–25 μ) | Pernicious anaemia |
| Anisocytosis | Much variation in size | Most anaemias |
| Poikilocytosis | Much variation in shape | Most anaemias |
| Spherocytic | Spherical | Congenital haemolytic anaemias |
| Target cell | Concentrically stained | Any chronic anaemia |
| Erythroblast | Nucleated | Denote excessive erythropoiesis |
| (1) Normoblast | Normal size | After haemorrhage, very severe |
| (2) Microblast | Small | anaemias and leukaemias |
| (3) Megaloblast | Large | Pernicious anaemia, carcinoma stomach, after total gastrectomy |
| Reticulocyte | Reticulated when stained with vital stains | If higher than 1 per cent in adults—an active marrow response to a demand for erythrocytes |

## Biopsy

Many lesions may be diagnosed only after examination of an appropriate biopsy specimen of the affected tissue. This is so, not only in cases of suspected neoplasia, but, for example, in the differential diagnosis of white patches which may occur in the oral mucosa and of the bullous, ulcerative, and desquamative lesions in the mouth. Many bone conditions are, similarly, only capable of final diagnosis by examination of a biopsy sample. It is generally agreed that, in the case of suspected or possible malignancy of the oral mucosa, biopsy is mandatory and, with simple precautions, is unlikely to cause dissemination of tumour cells. There are several methods of obtaining biopsies and these will be dealt with under the following headings: smear, excisional, incisional and biopsy for immunofluorescence.

### Smear biopsy

The examination of vaginal smears is well recognized as of great help in the diagnosis of early uterine and cervical carcinoma. Similarly, much progress has been made in the recognition of malignant cells, floating free in various body fluids, which originate from otherwise undetectable neoplasms.

This success in the recognition of desquamated tumour cells does not extend to malignant lesions of the oral epithelium. Smears taken directly from the surface of proved malignancies may well show only keratinized squames in the light microscope and give no indication of the nature of the lesion. Thus, it should be categorically stated that exfoliative cytology has no place in the routine diagnosis of suspected oral malignancy. It is always necessary to carry out biopsy of the excisional or incisional type and the accessibility of the oral mucosa makes this a relatively simple procedure. It is interesting to note that recent work has shown that scanning electron microscopy of cells exfoliated from oral malignancies may demonstrate characteristic morphological changes which may eventually be of use in diagnosis. However, this is as yet a completely undeveloped technique.

A study of epithelial smears may be of some value in the differential diagnosis of the bullous and other similar lesions of the oral mucosa (Fig. 1.12). The technique adopted is to scrape gently the surface of the affected area with a fairly broad square-ended instrument such as a spatula. (No preliminary cleansing of the area should be carried out.) The smear is transferred directly to a previously cleaned slide, spread into a thin film by the spatula and immediately fixed by dropping into 50 per cent ether-ethyl alcohol. If more than one area is to be studied, the spatula must either be carefully cleaned or replaced before the next smear is taken. If an intact bulla is available then the fluid contents should be aspirated and spread on the slide which is fixed in the same way. After fixing for a minimum of 10 minutes, the slides may be stained. For purposes of oral diagnosis, haematoxylin and eosin (H&E) is probably best, the complex stain specified by Papanicolaou for the study of cervical smears being somewhat less successful for oral conditions. A valuable alternative method of fixation is by the use of one of the commercially available spray products packed in pressurized containers and originally intended for use in cervical cytology. These are much more convenient than the highly inflammable ether-alcohol mixture.

Only positive results can be accepted in oral exfoliative cytology; essentially normal findings should be confirmed by further methods.

*Excisional biopsy*

If the lesion in question is small, it may be best to remove it entirely by local excision, including a small area of normal tissue. The specimen may then be sectioned and its histology reviewed to determine whether further treatment will be needed. The biopsy is far better taken with the knife than with the cutting diathermy which may cause considerable distortion of the tissues. If diathermy is felt necessary, as in a suspected malignancy, then the operation bed can be coagulated after the excision.

After its removal, the biopsy specimen should be placed with the minimum of delay into a fixative, 10 per cent formol saline being the most universally used. Full clinical details should always be given to the pathologist who is to study the specimen.

Excision biopsy is particularly useful for the diagnosis of single small ulcers and small localized soft tissue swellings. In these cases it is virtually equivalent to combining primary treatment with biopsy.

A special case occurs in the use of cryosurgery for the treatment of potentially malignant, highly vascular or similar lesions. The problem is that the whole of the tissue treated is lost by sloughing and is, thus, not available for histological study. In these circumstances a biopsy specimen may be removed from the area when frozen with little fear of haemorrhage or other troublesome complication. If it is thought essential, the whole of the frozen lesion can be removed in this way, combining some of the advantages of cryosurgery with those of biopsy excision.

## Incisional biopsy

This is the removal of a section of a lesion for histological study without any attempt being made to remove the whole of the lesion. In taking such a biopsy of the oral soft tissues, an attempt should be made to include within one specimen, if possible, a clinically typical area of the lesion and also the edge of the lesion. If the choice must be made between the two possibilities, the clinically typical area should be chosen; a large area of normal tissue beyond the lesion is quite unnecessary. The specimen should be big enough to give the pathologist a reasonable chance to make a diagnosis, as too small a biopsy is difficult to handle and to orientate for sectioning.

The technique for biopsy of a lesion of the oral epithelium, is to make a wedge-shaped cut into the chosen area, to complete the triangle by a third cut and to then take off the epithelial layer, together with a thickness of corium, by sliding the knife below and parallel with the surface. Even if it is the epithelium which is of particular interest, it is essential that a sufficient layer of corium should be included in order that the subepithelial reactions may be seen. If the biopsy is of a lump, then the wedge section must be taken into the swelling, making sure that any capsular tissue is cut through, and a representative area of the lesion proper is obtained.

Anaesthesia for the biopsy should be obtained by the injection of local anaesthetic as far from the biopsy site as consistent with obtaining a satisfactory result. It is clearly unwise to inject directly into an area of doubtful malignancy, and quite apart from any question of dissemination of neoplastic cells, there is a danger of distortion of the histological picture if the area is locally infiltrated by anaesthetic. The biopsy site may be closed by one or two silk sutures.

If the specimen is a thin one, as is often the case with biopsies of the oral mucosa, it is often most convenient to lay it flat on a piece of card or a swab before dropping into the fixative. The tissue practically always adheres to this backing and curling and distortion of the specimen is prevented.

Frozen sections are rarely required in oral medicine practice,; very often the specimens obtained by biopsy are in need of careful and detailed study, this being very difficult with frozen tissue. On rare occasions, however, when an urgent decision is essential on some major point (for example, whether a lesion is malignant or not) then a rapid result is obtained by cutting sections on a freezing microtome. Stained sections can be reported on in a very short time—of the order of a few minutes. The most usual form of apparatus depends on the cooling effect of a jet of carbon dioxide. On the whole, however, the sections are poor, the procedure is disliked by pathologists and it is advisable to ask for frozen sections only when essential, except in the case of immunofluorescent studies.

### Biopsy for immunofluorescence

Direct immunofluorescent studies of biopsy material have become an essential part of the diagnostic procedure for oral mucosal lesions (this is discussed further below and in the appropriate late chapters). In the case of non-bullous and non-erosive lesions there are no particular problems. The specimen is taken from the lesional tissue with some marginal clinically normal tissue if convenient. In the case of bullous or erosive lesions, however, the situation is quite different in that the most characteristic immunological findings are likely to be in the clinically normal tissue adjacent to the lesion. When bulla formation or erosion has occurred the technique of biopsy of the lesion itself becomes very difficult and the results (largely because of secondary infection and similar factors) become much more difficult to interpret.

It is essential when taking biopsy specimens for immunofluorescent studies that the laboratory should be aware of this and that the fresh unfixed tissue is passed on directly for immediate processing or for deep frozen storage.

## Microbiological investigations

If a bacterial aetiology is suspected for a lesion, both direct smears and swabs for culture and identification may be taken. Direct smears may be of some value in the identification of *Candida* and of Vincent's organisms, although their use is limited. In many instances only normal oral flora will be reported; this is the case, for example, in viral infections. However, in these circumstances, as in many other oral mucosal diseases, the balance of the oral flora is soon disturbed by the onset of abnormal environmental conditions.

*Candida* may be recognized on direct smears by culture and, if an estimate of density of organisms is required, by imprint culture. They survive well on a dry swab, but it must be remembered that in some forms of candidiasis, where the organisms are within the tissues (as in chronic hyperplastic candidiasis, see Chapters 3 and 7) there may be very little growth from a swab. They are best identified by histological methods. There are some other examples of this situation in oral lesions; for instance, in (the very rare) oral tuberculosis it may be necessary to culture fresh biopsy material in order to confirm the presence of organisms within the tissue.

Virus identification remains a lengthy and relatively difficult process. Immediate confirmation may be given by direct electron microscopy in the few centres where this is available. Otherwise, the only rapidly available evidence is by the presence of viral inclusion bodies in epithelial cells taken from the lesion by smear biopsy techniques. This is not a particularly accurate method of diagnosis and does not differentiate between the possible viral infections.

Antibody studies form the basis for accurate diagnosis of viral infections. The immune system reacts in two rather different ways which are perhaps best illustrated by examples. In *Herpes simplex* the baseline level of antibodies in any individual before clinical infection is variable, depending on the past history and the degree of the immune response. At the time of an active clinical infection these levels are raised considerably. If pairs of sera from the patient, taken at an interval (of about 10 days in this case) can be compared, a significant rise in titre confirms the diagnosis. Quite clearly, this is not a particularly useful technique, except in restrospect since, in the case of primary herpetic stomatitis, the lesions will have gone into remission before the confirmation of the diagnosis is available. In contrast to this situation, the individual who has not been exposed to the HIV virus will have no antibodies present. If infection takes place then the serum converts and antibodies are produced. In this and similar viral infections the presence of antibodies at any titre as a marker of exposure to the virus.

## Immunological tests

There are a wide range of immunological tests available to assist in the diagnosis of diseases affecting the oral cavity. With the introduction of much more specific immunological reagents than were previously available these have now become an essential part of the diagnostic processes of the oral medicine clinic. Many immunologically based tests are now matters of routine; others remain currently essentially experimental at the present time. For example, the use of monoclonal antibodies as cell surface characterizing agents in the assessment of premalignancy or malignancy, and in the investigation of the mechanisms of periodontal disease can, as yet, be considered only speculative. The use of monoclonal antibodies as highly specific targeting carriers of therapeutic agents is currently an even more tentative procedure.

Perhaps the widest ranging type of immunological screening test used in oral medicine is an estimate of circulating immunoglobulin fractions, complement components and other major immunologically active substances. A clue to the possible presence of immunoglobulin abnormalities may be given by the reporting of serum protein levels outside the normal range; this is a routine procedure and usually reported as part of the automated multi-channel blood analysis procedure now widely available. In autoimmune diseases, for example, total serum protein levels would be expected to be raised as a result of excess immunoglobulin production, whilst in immunodeficiency disease these levels would be expected to be reduced. Tests of circulating lymphocyte type and function are currently not so widely available.

The assay of specific circulating auto antibodies is an important procedure in the oral medicine clinic. Some autoantibodies are closely associated with specific disease processes—for instance, gastric parietel cell antibodies with pernicious anaemia. In other autoimmune diseases, however, a wide range of autoantibodies may be produced. For this reason it is usual to carry out a range of autoantibody tests rather than a single one. This also has the advantage of demonstrating the possible presence of more than one condition; for instance, anti thyroid antibodies are often associated with gastric parietel cell antibodies. Conversely, some autoantibodies are not disease-specific, but may be present in a number of conditions. The antinuclear antibodies, of various types, present in a number of connective tissue diseases are examples of this spectrum of association (Fig. 2.1).

The direct immunofluorescent technique for the detection of immunoglobulins and other immunologically active proteins fixed within tissue has acquired great importance in the diagnosis of oral mucosal lesions—particularly those associated with skin diseases and connective tissue disease. The principle of the technique depends on the fact that antibodies combined with fluorescein retain both their immunological activity and the property of the fluoroscein to fluoresce under ultraviolet light. Because of this the antibodies can be located at the exact site of combination with their antigenic antagonists by microscopic observation under ultra violet illumination. A wide range of highly specific antibodies to the various immunoglobulins and complement components is available and can be used to demonstrate the type and site of bound complexes. In some conditions the results may be highly specific and diagnostic (in pemphigus and pemphigoid, for example). In other conditions (such as lichen planus) they are less so. The findings in a number of conditions are discussed in Chapter 8. The rather special requirements for the taking of biopsy material for this technique have been mentioned above.

As has been pointed out above, there are many other laboratory- and clinically-based tests which give some indication of immunological abnormalities which may affect the oral cavity. These are only occasionally used at the present time although, quite clearly, in the currently rapidly progressing situation, this may change.

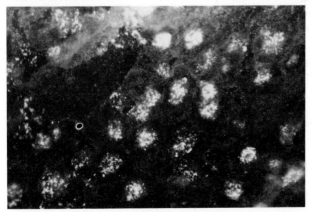

**Fig. 2.1.** Speckled type anti-nuclear antibodies demonstrated by immunofluorescent techniques.

A number of disease specific protein markers have been identified in the peripheral blood; some of these are readily available as slide tests in which only a drop of serum is required. These tests include the RA factor test in which the presence of a specific factor for rheumatoid arthritis is detected, LE factor for the detection of lupus erythematosus and C reactive protein. This last (CRP) appears in the serum of patients undergoing acute inflammatory reactions and in some other conditions such as rheumatoid arthritis. It is now generally accepted that these protein markers of disease are not particularly accurate and that the entirely non-specific erythrocyte sedimentation rate (ESR) is still probably the most useful generalized marker of disease which may be detected by changes in the plasma proteins.

Since oral medicine represents the whole field of medicine as related to the oral cavity, it is evident that as wide a range of investigations must be applied as in other specialities of medicine. It would, quite clearly, be impossible to even briefly outline these in the present context. The major investigative procedures have been outlined above. Others are discussed in the appropriate later chapters.

## Principles of treatment

In some lesions of the oral mucosa, specific treatment is directly indicated by the diagnosis. For instance, infections by an identified organism may be treated by the use of an appropriate antibiotic. However, in many oral lesions such rational therapy is not available, often because of a lack of knowledge of the factors causing the condition, and so it is necessary to fall back onto the treatment of the symptoms in order to give the patient relief. As has been discussed, many oral lesions present in similar forms in spite of differing aetiology and so a relatively restricted number of therapeutic agents may be of value in treating a wide range of conditions on a symptomatic basis.

A number of particularly helpful forms of symptomatic therapy may be considered. Details of specific drugs, dosages, and methods of administration will be dealt with in the appropriate later chapters.

### Covering agents

Some ulcers and erosions respond well to a simple non-active covering which eliminates, or reduces, secondary infection and irritation. A number of gels and pastes are available for this purpose—no more permanent covering agent has as yet found general use. Such agents as carboxymethylcellulose paste may be used as carriers for more active substances such as steroids or antifungal agents. A number of preparations of this kind are commercially available.

### Antibiotics

Apart from the use of antibiotics and other chemotherapeutic agents by the usual systemic route, the use of antibiotics locally is justifiable in certain conditions. It goes without saying that there are inherent disadvantages, particularly of the possibility of resistant organism production and of inducing hypersensitivity reactions in the patient, but the value of the treatment is such as to make such risks justifiable in some cases. Tetracycline (or chlortetracycline) is the most useful antibiotic used in this way. As a 2 per cent solution it is often effective in reducing the secondary infection (and thus the pain) in cases of aphthous ulceration, herpetic stomatitis, erosive lichen planus, and other severe ulcerative conditions. However, it must be stressed that this cannot normally be accepted as treatment for recurrent conditions and it is best to regard it as emergency treatment reserved for acute episodes. The mouthwash may be made by the patient dissolving the contents of one 250 mg capsule in 15 ml of water to given an approximately 2 per cent solution. It is perhaps better to have the solution accurately made up by the pharmacist, including 10 per cent of glycerol as a demulcent. If the treatment is not unduly prolonged, there is minimal trouble from overgrowth of resistant organisms in the mouth although very occasionally a fungal infection may occur and must be appropriately dealt with. Many of the author's patients have oral lesions which are persistant and severe. In such cases, the prolonged use of antibiotic-based mouthwashes is clinically justified, particularly in the chlortetracycline–triamcinolone combination (detailed below) which is used in such conditions as pemphigus, major erosive lichen planus, and other similar conditions. Oddly enough, chlortetracycline mouthwashes are almost specifically active in reducing the discomfort of herpetiform ulceration (see Chapter 4). It is particularly difficult to refuse repeated prescriptions for antibiotic mouthwashes in the case of recurrent oral ulcerative conditions in which this form of therapy has been found effective.

There are a number of antiseptic mouthwashes available which are of value in reducing secondary infection and, hence, painful symptoms in many conditions. On the whole these are less effective than tetracycline mouthwashes, but do not have the limitations of an antibiotic, especially in terms of repeated usage. Chlorhexidine, benzydamine, and hexitidine preparations are the most often used. It is often difficult to predict which of these will have the maximum effect in any given situation. There is a considerable factor of patient preference to be considered. There is little evidence that any one is superior to the others when used for this generalized purpose.

## Steroids

Locally applied corticosteroids are effective in the control of certain conditions such as aphthous ulceration and erosive lichen planus. The steroids have, apparently, two modes of action. In the first place there is a generalized anti-inflammatory response, whatever the basic aetiology of the ulceration. In conditions which seem to depend on an autoimmune mechanism there is also a specific response to steroids which is quite separate from, and more powerful than, the simple anti-inflammatory response. In fact, one of the criteria of an autoimmune process is taken to be its sensitivity to steroids. One of the most important factors to be considered when using steroids is the degree of suppression of adrenal function which may occur when these drugs are taken. The degree of adrenal suppression varies not only from steroid to steroid and according to the method of use, but also in response to individual variation. For instance, a dose of systemic prednisolone which may apparently cause no side effects in one patient may render another markedly Cushingoid. In general, steroid responsive oral lesions (such, for example, as those of pemphigus or in Beçhet's syndrome) require high dosages to achieve a satisfactory effect—often higher than those needed to suppress equivalent skin lesions. For reasons such as these it is the authors' practice to use high concentration, locally applied steroids to replace systemic medication whenever possible.

Application of corticosteroids locally may be by means of lozenges, in mouthwashes, in creams and gels, and by aerosol spray. These are available in a number of commercial products which err on the side of caution in terms of possible side effects as outlined in the product information sheets. All of the available products designed for intraoral use, when used in the recommended manner, are insufficiently active to cause adrenal suppression. However, it is unfortunately true that they are often also insufficiently active to make much impression on the condition for which they are used. The most useful steroid aerosol in the author's hands has been a betamethasone valerate aerosol, originally designed for asthma therapy, which can be easily adapted for oral use.

The use of systemic steroids for the treatment of lesions of the oral mucosa is outside the province of normal dental practice. Although a small proportion of ulcerative and bullous lesions may justify such treatment, careful assessment and hospital supervision are essential. With the necessary precautions, however, favourable results may be obtained in some obstinate cases by the use of short-term courses of steroid treatment. Occasionally, a course of prednisolone of a week's duration, carefully tailed off, may greatly help a resistant case of oral ulceration and will have no untoward effects. Rarely, an intractable and severe case may justify and necessitate continuous systemic steroid therapy, with its associated disadvantages and risks, but in such patients the oral symptoms are usually no more than part of a widespread and more generalized condition.

As an alternative to the use of systemic steroids, high concentration steroid mouthwashes may be used. These (in dosages determined by the precise clinical situation) may be very helpful, but it should be realized that systemic absorption and consequent side effects may occur. The author uses triamcinolone mouth-washes, most often in a chlortetracycline base, at a total dosage which can be varied according to the severity of the condition. Mouthwashes containing levels of triamcinolone varying from 0.75–1.5 mg to 3–6 mg daily are most commonly used.

Intralesional injections of depot preparations of suitable steroids, often used by dermatologists to produce localized high concentrations of the active drug, may be used to treat oral lesions, but are, in general, not so effective as in dermatological practice.

## Anaesthetics

Very occasionally it may be in the interest of the patient with a particularly severe oral ulceration or erosion to prescribe a local anaesthetic preparation in order to enable food to be taken. It is quite evident that this is simply symptomatic treatment and that it is a manoeuvre which should be approached with caution since secondary trauma could easily be painlessly effected during the period of anaesthesia. A further precaution to be taken includes the avoidance of preparations of sufficient strength to affect the laryngeal reflexes. However, in spite of these difficulties it is, from time to time, justifiable to prescribe treatment of this kind. A wide range of solutions, lozenges, and sprays is available.

## Artificial saliva

The problem of the dry mouth is a particularly difficult one to solve—there is discussion of this in Chapter 5. There is a wide range of available saliva substitutes, some containing glycerol, some methylcellulose, and some mucins as the demulcent agent. They are presented either as a simple mouthwash or as sprays. In some formulations there are buffering agents and other ingredients meant to simulate the ionic composition of saliva. Others contain fluoride in order to combat the aggressive caries which may occur in the very dry mouth, particularly after radiotherapy. All of them, however, suffer from the disadvantage of being a temporary measure; the effect is relatively transient and it is necessary for the patient to repeatedly use the spray or mouthrinse in order to have any continuing effect.

Many other therapeutic agents are of use in the practice of oral medicine, these will be mentioned in the appropriate chapters. It should always be remembered, however, that the symptomatic treatment of oral lesions should be secondary to the systemic investigation and diagnosis of the basic disease process.

# 3

# Infections of the oral mucosa

It is possible to demonstrate a wide range of potentially pathogenic organisms in the normal mouth. Many of these are present in relatively high concentrations, but in spite of this oral mucosa shows a remarkably low susceptibility to primary infection. This is probably due, at least in part, to the activity of the saliva, from which a number of antibacterial substances may be separated. As has been pointed out in Chapter 1, IgA is also secreted in the saliva and it has been suggested that, as well as a possible direct effect of this on the oral organisms, a complex may be formed between the immunoglobulin and the oral epithelium itself resulting in a protective surface coating on the mucosa. There is also a further mechanical protective activity of the saliva in that foreign material, including organisms, is washed from the oral cavity to the stomach where bacteria are destroyed by gastric fluids. This relative insusceptibility to infection and inflammatory changes of the oral mucosa in general does not extend to the crevicular tissues of the gingivae. It has been demonstrated that macromolecules and relatively large particles of solid material may pass through the thin crevicular epithelial barrier into the tissues below and, although it is not certain that bacteria are able to pass through the epithelium in this way, the local microulcerations which are the end result of this partial permeability provide an easy pathway of ingress for organisms of all kinds. The complex immunological factors operating in the region of the gingival crevice have been outlined in Chapter 1.

However, these local factors represent no more than a first line of defence for the patient against invasion by potential pathogens, and depend for their effectiveness on the integrity of the immune and other generalized protective responses of the host. If the balance between the host and the commensal organisms is disturbed by some factor which impairs the immune defences then the organisms may begin to act in a pathogenic manner. It is in such a way that many cases of oral candidiasis and acute ulcerative gingivitis are thought to occur. Apart from infections of the oral mucosa brought about by a disturbance of a normal host–commensal relationship there are others, particularly those of viral origin, which represent the first response of the patient to the infective agent. Even in some of these (orofacial herpes infections are an example) there may be complex immunological changes in the patient which are of great significance in understanding the clinical course of the disease.

# Acute ulcerative gingivitis (acute necrotizing gingivitis)

This condition is the most widespread of the superficial infections of the oral mucosa and the most likely to be seen in general dental practice. Although acute ulcerative gingivitis (AUG) is such a common condition its exact cause is not known. There is no doubt that during an attack there is a great proliferation of the Vincent's organisms (*Borrelia vincenti* and *Fusiformis fusiformis*) within the affected tissue, but there are several ways in which it appears that simple infection with these bacteria is not the initiating factor in the disease. However, although there are many theoretical speculations about the exact aetiology of AUG there is no doubt that for practical purposes elimination of the overgrowth of the organisms is coincident with clinical remission of the disease.

The usual course of the condition may be summarized as follows:

(1) soreness and bleeding of the gingivae;

(2) the development of crater like ulcers initially involving the gingival papillae;

(3) lateral spread of the ulceration along the gingival margins (Fig. 3.1).

The ulcers have a grey–white base with a linear red border and are accompanied by a marked halitosis which is generally accepted as being characteristic of the condition, but which is considered by some authorities to be no more than a sign of generalized oral stagnation. It has also been suggested that the fever, malaise, and lymphadenitis developed by some patients is not specific, but that it is more suggestive of a mixed viral and bacterial infection. However, whilst there can be little doubt that a viral stomatitis may produce these symptoms there is equally no doubt that they are also produced in patients in whom, by all clinical and bacteriological criteria, AUG is the proper diagnosis.

Most patients with AUG are young adults and there is a marked seasonal variation in the incidence of the condition, the greatest incidence in Europe

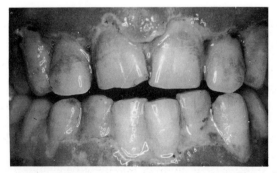

**Fig. 3.1.** Acute ulcerative gingivitis showing destruction of papillae and lateral spread of ulceration along gingival margins.

being in the winter months. In the more usual, uncomplicated, case of AUG the patient is initially perfectly healthy. The influence of poor oral hygiene in the initiation of this condition is often stressed, but there is no doubt that there are some patients with AUG in whom the previously maintained standard of hygiene must be considered by normal criteria to be good. Further reputed predisposing factors are tobacco smoking and psychological stress, although it is by no means easy to see how these factors operate. So far as smoking is concerned it has been suggested that a possible mechanism is that excess adrenaline may be liberated following stimulation of the adrenals by nicotine. Vasoconstriction in the vessels of the gingivae may follow, with a consequent reduction of blood supply to the tissues and increased susceptibility to infection and damage. It is also suggested that there may be an enhancement of the vasoconstrictive effect in the presence of endotoxins from Gram-negative bacteria in the gingival plaque nearby. This rather surprising theory gains support from the established patterns of infection, which on a local scale appear to be closely associated with the distribution of the gingival blood vessels.

As has been pointed out, the bacteriological aspects of AUG are interesting and unusual. During an attack there is invariably an overgrowth of the two Vincent's organisms (*Borrelia vincenti* and *Fusiformis fusiformis*). However, a gingival smear taken during an attack usually shows the presence of a wide range of other organisms as well as these two. The Vincent's organisms are present as commensals in many normal mouths and there are considerable doubts as to whether the disease is due to a primary infection by them or whether their proliferation is secondary to some other process, the true nature of which is not understood. The most important reason for the doubts regarding the aetiology of AUG is that the classical criteria, first specified by Koch to define proof of bacterial infection and since universally accepted, do not apply to this disease. The 'Koch's postulates' are as follows:

(1) the organisms involved must be present in every case of the disease;
(2) the organisms must be capable of in-vitro cultivation over several generations;
(3) the disease must be transmissible in its characteristic form following inoculation by the organisms concerned.

It is this third factor which so far eludes verification in the case of AUG. Many attempts have been made to transmit the disease by direct inoculation of human volunteers with the organisms, but the results have been almost uniformly unsuccessful, with the exception of a very few instances in which the inoculated subject was already in a state of ill health, with a cold or having recently had influenza. Experimental lesions in animals have proved very difficult to demonstrate and none have been induced by the inoculation of the Vincent's organisms alone. Those experiments which have resulted in the production of what are described as typical fusospirochaetal infections have all involved the use of a number of organisms, usually four, a fact which is compatible with the wide range of organisms found in smears from the gingivae in AUG. However, the induction of a lesion in the guinea-pig groin is a situation rather remote from the clinical conditions at issue and, although the work in question raises many points of great interest, it cannot be said that the essential bacteriological basis of AUG is as yet elucidated by such experiments. In general terms, it seems reasonable to accept that organisms other than Vincent's organisms may play a part in the initiation of AUG by providing suitable environmental conditions for the proliferation of the fusobacteria and spirochaetes, although the exact nature of the role of these other organisms is not clear. In spite of the considerable theoretical interest in the true origins of AUG and in the function of the organisms involved there is, so far as therapy is concerned, little to be lost by considering it as a simple infection by the Vincent's organisms and by aiming treatment towards elimination (or at least a drastic reduction) of them.

In normal healthy patients with AUG spread of the infection from the gingival margins is relatively rare although, very occasionally, contact ulceration of the buccal mucosa may be seen, but in a patient weakened by debilitating disease (as for instance leukaemia) the infection may become rampant. An example of the spread of AUG in the debilitated patient is in cancrum oris, a condition virtually unknown in European conditions of nutrition at the present time, but relatively common in earlier times. It has been shown that the rapid and massive destruction of the facial tissues which occurs in some African children is due to a combination of suppression of the immune system by malnutrition, a lack of oral hygiene, and the possible initiating stimulus of a viral infection. Under these circumstances the condition, presenting initially as quite typical AUG, spreads with extreme rapidity and, within a matter of a few days widespread soft tissue necrosis may occur. At this stage, provided that the patient does not die of toxaemia or of secondary effects, the infective process often comes under control, presumably by belated operation of an immune response.

One of the greatest clinical problems in the management of AUG is that of recurrence. It is evident that the patient with poor oral hygiene or with gingival contours distorted by a previous attack may, because of these local factors, be susceptible to recurrent infection and it is equally evident that a patient with underlying systemic disease may also be liable to recurrence. It should be stressed that any patient who, for no obvious reason, has recurrent attacks of AUG should be suspected of suffering from an underlying systemic disorder and, in view of the preponderance of blood dyscrasias among such patients, blood examination is a mandatory part of the screening process. Apart from such patients with evident local or systemic abnormalities there remains a further group in which no such predisposing factors are present, but in whom AUG remains persistently recurrent. In an investigation of a group of such patients Lehner estimated the serum immunoglobulins during and after the attacks, and found that there was depression of serum IgG in all patients with AUG, this being most marked in those subject to recurrence. It was also found that IgM was significantly raised both in those patients undergoing a first attack and in those patients with recurrent attacks, whilst IgA was found to be normal in patients undergoing their first attack, but was significantly depressed in the recurrent patients.

This pattern of immunoglobulin distribution is somewhat surprising since IgG levels are usually raised in bacterial infections. It is pointed out by Lehner that there are two possible explanations: either the existence of an intrinsic IgG deficiency or a suppression of IgG synthesis by endotoxin production during the early phase of the disease. Since the Gram-negative organisms which produce such endotoxins are present in abundance as part of the flora of patients with poor oral hygiene this is a suggestion which fits neatly with the bacteriological observations previously discussed.

It is suggested by the same investigator that the suppression of IgA in recurrent cases may follow a similar pattern to that of IgG, following an attack there is a rise towards normal values of both IgG and IgA, this coinciding with the reduction of the numbers of Gram-negative organisms in the area. An increase in IgM, as shown in AUG is commonly associated with either viral, protozoal, or Gram-negative infections. There is an interesting association here in the activity of the drug Metranidazole not only against AUG, but also against trichomoniasis and amoebiasis which are protozoal infections. This is no more than a speculative association although a quite high incidence of amoebae (60 per cent) and trichomonas (12 per cent) has been found in gingival smears from AUG. The association of cancrum oris with viral diseases has previously been mentioned and it is interesting to note that in viral infections the common immunological changes are suppression of IgG and elevation of IgM, precisely the changes shown in AUG. Detailed immunoglobulin estimations have not been carried out on patients with cancrum oris, but it may well be that virus-induced immunological changes of this type may be responsible for the lack of immune control in these patients. In summary, this important work by Lehner leads to a conclusion that an immune abnormality may exist in patients with AUG. In a single attack this abnormality may result from local factors such as the presence of endotoxin producing organisms, whilst in the case of patients with recurrent attacks the abnormalities may represent a permanent immune defect.

The immediate treatment of AUG depends on the elimination of the overgrowth of Vincent's organisms. At the same time the treatment must be aimed towards the restoration of a gingival condition in which circumstances favourable to a recurrent attack are not present. It is evident that any coexistent potentiating systemic disease must also be treated. Whatever the circumstances, however, the local treatment depends on three possible lines of action:

1. The elimination, by oral hygiene measures, of any stagnation areas in which further proliferation of the anaerobic Vincent's organisms might take place. Many periodontists insist that little other than scaling and polishing is

necessary in the treatment of the majority of cases, but the standard of care required is very high and the time expended on treatment may be unduly long if this approach is adopted. However, it is clear that the attainment and subsequent maintenance of a high standard of oral hygiene is the key factor in the prevention of recurrence of the condition.

2. The use of oxygenating agents. In the traditional treatment of AUG these substances, particularly chromic acid and hydrogen peroxide, played a leading role. They are now considered to be of secondary importance, but are accepted as being of value as adjuncts to treatment, the exception being chromic acid which is now looked upon with great disfavour as it is itself capable of causing considerable tissue destruction. However, the use of dilute solutions of hydrogen peroxide as a mouthwash (10 vol. solution diluted four or five times is often used), or buffered sodium perborate mouthwash are acceptable aids to treatment. It has, in fact, been reported that either of these preparations, if carefully used, may produce results as satisfactory as those following chemotherapy. The value of oxygen-liberating mouthwashes and applications has long been thought to be due to the conversion of anaerobic conditions to aerobic with consequent discouragement of the Vincent's organisms, but it is suggested by some authorities that the amount of oxygen released during the short period of exposure is relatively insignificant and that the greater value of the oxygenating substances is as mechanical cleansing agents, removing debris by effervescence.

3. Chemotherapy: AUG responds rapidly to the use of penicillin, tetracyclines, and a number of other antibiotics. If, for any reason, this form of treatment is chosen, a normal systemic dosage should be used and local preparations avoided. However, there is now rarely any justification for the use of these antibiotics since Metranidazole is readily available. It has been shown quite clearly that this drug, in the vast majority of cases, is equal in efficiency to penicillin in the treatment of AUG and there are, as yet, no problems of organism resistance or of patient hypersensitivity. A dosage of 200 mg, three times daily, is sufficient to reduce the symptoms dramatically in 24 hours. However, it must be stressed that treatment of AUG by Metranidazole or by antibiotics is no more than a preliminary procedure which must be followed by meticulous oral hygiene treatment. It has been suggested that Metranidazole should not be used in patients with blood dyscrasias and all patients taking the drug should be warned of the effects of alcohol. Metranidazole, like the drug Antabuse (disulphiram) blocks the normal metabolic pathway of alcohol elimination and, in some patients, side effects such as nausea may follow if alcohol is taken during treatment. Like other drugs, it is not recommended in early pregnancy.

## Streptococcal stomatitis

There would seem to be no doubt that there is a readily available reservoir of streptococci in the oral cavity which might be expected to respond to a disturbance in the host–commensal relationship by behaving in a pathogenic manner. It is equally evident that a generalized infection of the mouth and oropharynx might occur as a result of invasion by exogenous cocci. In spite of this it is by no means clear that a true streptococcal stomatitis does, in fact, exist although until relatively recently it was generally accepted as a clinical entity. The confusion arises because of the similarity between the symptoms ascribed to a streptococcal stomatitis and those caused by mild viral infections. The symptoms of streptococcal stomatitis are generally described as a generalized erythema of the oral mucosa together with a marked gingivitis, submandibular lymphadenitis, and a mild degree of malaise and fever. There is no doubt that in some cases of stomatitis corresponding to this description a heavy growth of streptococci can be obtained from the mouth, but it is equally true that precisely similar populations of cocci may be grown from patients in whom the stomatitis would seem on all grounds, including the evidence obtained from serum antibody studies, to be viral in origin and in which the streptococci would appear to be playing the part of an opportunist secondary invader.

Although the difference between a primary streptococcal stomatitis and a secondarily infected viral stomatitis is of evident theoretical interest the possibility of confusion introduces no real complication into the treatment of the condition. Should the stomatitis be mild, then no more active treatment than, possibly, a bland mouthwash to help maintain oral hygiene is required. If the condition is more severe the use of 2 per cent tetracycline mouthwash is effective in reducing discomfort and the length of time for which symptoms are present. Systemic antibiotic therapy is likely to be no more effective, and may be less so, than local therapy of this kind.

## Syphilis

The number of primary oral lesions of syphilis seen in an oral medicine clinic is very small. However, primary lesions may occur on any part of the oral mucosa or lips and, in spite of their relative rarity, must always be considered in a differential diagnosis of oral ulceration. As in the case of the genital lesion the oral primary lesion (chancre) appears following a period of 2–3 weeks after infection. Usually the lesion consists of a painless indurated swelling, dark red in colour and with a glazed surface, from which large numbers of *Treponoema pallidum* can be isolated (Fig. 3.2). This isolated and discrete lesion is most likely to be found in the relatively soft and unrestricted tissues of the tongue, cheeks, or lips, but if the lesion occurs on the palate or gingivae the morphology of the lesion is modified, and it may appear as a more diffuse and much less discrete structure. Whatever the site, size, or shape of the chancre, however, the heavy infection of the surface is consistent, this providing both a convenient means of diagnosis and a considerable hazard to the unsuspecting diagnostician. At the time of the primary lesion there is a non-tender enlargement of the cervical lymph nodes affecting the submental, submandibular, pre-, and post-auricular

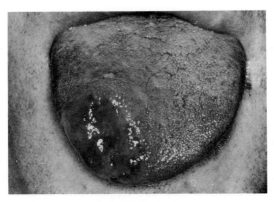

Fig. 3.2. Primary syphilitic lesion (chancre) of the tongue.

and occipital groups—the so called syphilitic collar. These nodes also contain a large number of proliferating spirochaetes; aspiration of these is possible as a confirmatory diagnostic measure.

Diagnosis at this early stage is largely on clinical grounds, confirmed by examination of material taken from the surface of the chancre or aspirated from a cervical node. The most satisfactory form of examination is immediate dark-ground microscopy and, by this technique, the spirochaetes can be identified quickly and an immediate diagnosis made. However, this form of examination must be carried out by an expert; there are spirochaetes present in the normal oral flora which might easily be mistaken for *Tr. pallidum*. It should be remembered that all the serological tests for syphilis may give negative results in the very early stages of the disease. The most sensitive test in the primary stage is the fluorescent treponaemal antibody [FTA(ABS)] test, which is positive in about 90 per cent of patients. The VDRL test (the Venereal Disease Research Laboratory test) is positive in about 75 per cent of primary cases. In the secondary stage these tests approach 100 per cent accuracy although false positive results may be given in some other tropical treponaemal diseases.

This disappearance of the primary lesion, usually after a period of some 2 weeks, marks the widespread dissemination of the organisms and the onset of the second stage of the disease which may last for many years. The oral symptoms most often described at this stage of the disease are mucous patches (appearing as grey–white ulcers covered by a thick slough) and snail track ulcers, and cases which conform to these descriptions have often been described. However, in view of the protean nature of the skin lesions produced during this stage of the disease it would seem at least a reasonable possibility that there may be an equivalent variation in the form of oral lesions. If this is so it would seem more than likely that secondary syphilitic ulcers may often pass unrecognized or be mistaken for some less significant non-specific lesion and for this reason serological tests for syphilis should be part of the investigation of all oral ulceration of unknown origin. In this second stage lymph nodes may again be palpable as non-tender, discrete structures.

In the tertiary stage of syphilis two major forms of oral involvement may occur. The first of these, syphilitic leukoplakia, will be dealt with in detail in Chapter 4 together with other leukoplakias. For the moment it is sufficient to point out that, in the inadequately treated or untreated cases which are now seen very rarely, leukoplakia of the whole of the oral mucosa, but especially of the tongue is an important complication, the more so because of its reputation for malignant transformation. The second likely form of oral involvement in the tertiary stage is in the tissue destruction consequent on gumma formation. The gumma, essentially a chronic granuloma, forms as a solid structure, often in the palatal tissues, which eventually breaks down with the consequent production of a tissue defect (Fig. 3.3). The untreated patient at this stage is likely to have other more widespread lesions, especially of the nervous system, but the organisms are by no means as readily demonstrated as in the earlier stages of the diseases. As well as these most commonly described manifestations of late syphilis a number of other oral changes have been described, including a fibrosing glossitis and angular cheilitis. It is evident that a wide range of clinical presentations of tertiary syphilis is possible and this fact emphasizes the wisdom of including serological tests for the disease in the investigation of unusual oral conditions. Even if the patient is of an advanced age when the condition is diagnosed (as in the patient shown in Fig. 3.3) and although no systemic effects are obvious, treatment is indicated. Symptoms such as mental confusion which might be attributed to senile changes may be, in fact, the result of the syphilitic infection and may respond to antibiotic treatment even at a late stage.

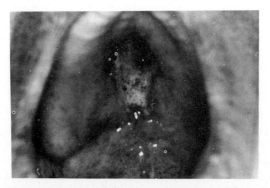

**Fig. 3.3.** A tertiary syphilitic lesion (gumma) of the palate.

# Gonorrhoea

Primary oral lesions of gonorrhoea are relatively rare. They are thought to be, in the majority of cases, the result of transmission of the organism (*Neisseria gonorrhoeae*) by direct mucosal contact. Purulent gingivitis, diffuse erythematous and ulcerative lesions, tonsilitis, and other oral manifestations have been described. However, patients demonstrated to have oral lesions are very few relative to the very large number known to have genital gonorrhoea. The systemic disturbances described as being associated with such oral lesions vary from mild to severe febrile symptoms and the degree of oral discomfort reported is equally variable. It is likely that in most cases, the nature of the infection is not initially suspected on clinical grounds, but becomes clear only as a result of bacteriological examination. It may well be that, because of the relatively non-specific nature of the lesions, the true incidence of oral gonorrhoeal infections remains unrecognized.

A diffuse form of gonorrhoea, spread by haematogenous routes, may very occasionally affect the oral mucosa although the predominant manifestations are of the skin. Ulcers, haemorrhagic lesions, and other manifestations of hypersensitivity to the disseminating organisms have been described. Occasional patients with gonococcal infective arthritis of the temporomandibular joints have been reported. The symptoms are as would be expected from an infective arthritis—pain, swelling, and trismus. In most described cases the diagnosis has depended on bacteriological study of fluid aspirated from the affected joint. This illustrates the fact that, in the absence of an effective immunological test, gonorrhoea in the oro-facial area is likely to be under diagnosed.

As in genital gonorrhoea, the treatment is by high doses of antibiotics; varying regimes have been described varying from single high dose penicillin injections (4.8 million units aqueous procaine penicillin G + 1 g oral probenicid) to oral penicillin (3 g amoxicillin + 1 g probenicid) to longer term courses (2 g tetracycline daily for 7 days). It is generally accepted that oral gonorrhoea needs relatively vigorous treatment in comparison with genital, presumably as a result of relatively poor antibiotic secretion in the saliva and equally poor transmission across the oral mucous membrane. As in all venereal diseases it should be remembered that the patient might have acquired more than one infection at the same time; serological tests for syphilis are mandatory.

# Non-specific urethritis: Reiter's syndrome

Non-specific urethritis (NSU) is probably the most common sexually transmitted disease. Its aetiology is not clear although in many cases *Chlamydia* species may be isolated from the urethra. The symptoms are of burning sensations on micturition. The majority of cases respond to tetracycline therapy in normal doses over a week or two, even though no microbial agent can be demonstrated.

A very few of these patients (usually young men) develop Reiter's syndrome.

In this condition there may be a polyarthritis (which may affect the temporo-mandibular joints), conjunctivitis, and oral lesions resembling geographic tongue which may also affect other areas of the oral mucosa (circinate stomatitis). The great majority of patients affected by Reiter's syndrome carry the HLA B27 antigen.

## Tuberculosis

Tuberculosis was in the past a common cause of oral ulceration, the oral mucosa becoming secondarily infected by the sputum in cases of active pulmonary disease. However, with the virtual elimination of such cases the number of oral ulcers attributable to this cause has dwindled almost to vanishing point. None the less, in the very occasional case of a persistent oral ulcer of unknown origin, a tubercular origin must be considered in the differential diagnosis. The classical description of the tuberculous ulcer is of an irregular lesion with undermined borders and covered by a grey–yellow slough. Diagnosis of such a lesion is likely to follow a biopsy, and a caseating tuberculoid pattern in the underlying tissue would be highly significant. However, other chronic granulomatous lesions (as, for instance, in Crohn's disease) although non-caseating, may be very similar in their histology. The final diagnosis is best made by culture of fresh biopsy tissue to isolate the organism. The number of organisms present in oral lesions may be very low and their demonstration by Zahl–Nielsen staining or by immunofluor-escent techniques may be unreliable. It need hardly be said that any patient should be fully investigated for pulmonary or other lesions and, indeed, it would be expected that such measures as a chest X-ray would be taken at the first moment of suspicion.

It should be added that a few lesions have been described in which the presentation of tuberculosis of the oral mucosa is quite different from the ulcers as usually described. These lesions, presenting as white patches or granulating lesions, have been described both as primary lesions and as lesions secondary to pulmonary infection.

Clearly, the treatment of tuberculosis is outside the province of oral medicine and the appropriate action to be taken would usually involve reference of the patient to a physician experienced in the treatment of the more common forms of the disease. The treatment regimes adopted conform to those used for pulmonary disease.

## Candidiasis

Candidas are fungi which have a wide distribution and which frequently form part of the commensal flora of the human body. Swabs taken from the skin, gut, vagina, or mouth of an apparently healthy individual all may show the presence of *Candida* species and, in particular, the form *Candida albicans*. It has long been recognized that disease due to a proliferation of this organism is a mark of lowered resistance or of metabolic change in the patient and that the onset of candidiasis must lead to a search for the underlying cause. Some of these predisposing conditions are:

(1) local tissue trauma (this is a particularly important factor in oral candidiasis);

(2) endocrine disorders;

(3) malnutrition and malabsorption states;

(4) patients in whom the normal bacterial flora is unbalanced by antibiotic treatment;

(5) blood diseases;

(6) patients weakened by malignant disease or by surgery;

(7) patients in whom the immune response is defective or suppressed by drugs;

(8) temporary physiological weakness, as in infancy or pregnancy.

It should be remembered that the onset of candidiasis in an adult patient represents a change implying a relaxation of the normal immune defences, it does not imply an infection by 'foreign' *Candida* strains. Although, in the present context, it is evident that the oral conditions involving *Candida* are those more fully discussed it must be remembered that candidal lesions may involve other areas and, indeed, may become widespread and disseminated in a few debilitated patients. The condition septicaemic candidiasis may involve the lungs, myocardium, and other vital organs, and is always a disease of very poor prognosis. Apart from such widespread infections, however, infection of the skin, hair follicles, and nails is relatively common. As in the case of oral candidiasis local tissue trauma may be a significant predisposing feature and, in the case of candidal paronychia, circulatory disturbances and diabetes are commonly the precipitating factors. There is some difference of opinion as to the incidence of *Candida* in the vagina, most investigations have shown a relatively low incidence in non-pregnant patients, but a high one (of the order of 50 per cent) in patients during late pregnancy. Not all these patients by any means show evidence of clinical infection, but it is evident that this vaginal population of *Candida* must act as a reservoir for infection, particularly of the new-born baby.

*Candida albicans* normally exists in the oral cavity in the form of rather large yeast-like cells which occasionally elongate to form pseudohyphae or germ tubes. In the inactive state the yeast form is predominant, but when pathological activity occurs the hyphal form is much more evident. These pseudohyphae can be seen, not only superficially on the oral mucosa, but also penetrating deeply into the epithelium. Electron microscopic studies of the organisms in this invasive form show the presence of a very large number of active organelles, a mark of increased cellular activity. Isolated strains of *Candida* do not vary greatly in their pathogenic effect in experimental animals, and organisms from a wide variety of sources show similar pathogenicity. It has been shown in experimental conditions that a significant factor in the growth of *Candida* is the influence of relative humidity, but it is not always possible to directly relate *in vitro* growth characteristics with clinical behaviour. The mechanism by which *Candida* exert a pathological effect on tissues is not known and, although it has been quite clearly demonstrated that a toxic substance may be produced by the organisms and that this can induce skin lesions in the absence of the organisms themselves, it is not clear whether this is a substance which acts in the same way as a bacterial toxin. There is, in fact, some evidence that the toxic product may act as an antigen, stimulating an allergic response in the affected tissues.

The predisposing factors to oral candidiasis may be summarized as follows.

1. *Intrinsic immune defects.* These may occur in either the cell-mediated or humoral mechanisms. Both play a part in the immune control of *Candida* and, if one or other of these factors is deficient, chronic candidal infection is not unusual. As was pointed out in Chapter 1 such patients are rare and, indeed, many die at an early age as a result of infection. However, in those who survive, chronic candidiasis affecting mucous membranes and skin may prove an intractable problem.

2. *Patients undergoing immunosuppressive therapy.* There are many more of these patients than in the previous group since many patients are now under treatment by drugs which induce at least some degree of immunosuppression. A number of drugs exert such an effect on the immune response. Of these perhaps the most generally used are the corticosteroids which suppress both the cell-mediated and humoral responses. Other drugs which may have an immunosuppressive effect include cytotoxic drugs (such as busulphan and cyclophosphamide) and other drugs which affect cell metabolism (such as purine analogues and folic acid antagonists). The immunosuppression may be an entirely unwanted side effect (as in the cytotoxic drug treatment of neoplasms), but in other circumstances the immunosuppression may be deliberate. An example of this is in the treatment of patients undergoing transplant surgery in whom the immunosuppressive therapy is aimed at the elimination of rejection phenomena, Similarly, immunosuppressive therapy may be of great value in the treatment of diseases of autoimmune origin. In such instances corticosteroids are often used to suppress the immune response as a whole and this includes the harmful autoimmune process. It is unfortunately true, however, that such immunosuppression is entirely non-selective and the patient may well become particularly susceptible to infection of all kinds as a result of the loss of the normal immune response.

3. *Patients on antibiotic therapy.* The disturbance of the normal balance of the oral flora in patients undergoing antibiotic therapy may result in an overgrowth of *Candida* in the abnormal environment. This, however, is no more than an occasional problem in relation to the large numbers of courses of antibiotic

treatment prescribed. However, the development of the 'antibiotic sore tongue' in a patient on such therapy represents something of a problem since this rather painful superficial candidal glossitis is often slow to clear. This condition will be further discussed below.

4. *Generalized disease processes.* These may include a very wide range of conditions including blood diseases of various kinds and endocrine disturbances. Iron deficiency seems to be of particular importance in this respect. It has been pointed out that not only those patients with evident anaemia as shown by changes in haemoglobin concentration and erythrocyte morphology, but also those with none of these changes, but with reduced serum iron may be particularly susceptible to candidal infection. As has been pointed out in Chapter 1, a full range of haematological investigations, including serum iron, folate and B12, and red cell folate estimations, should be carried out as part of the routine investigation of any patient with candidiasis which is not of evident aetiology.

However, one of the most significant conditions which may lead to the onset of persistant oral candidiasis is AIDS. Otherwise unexplained candidal infections are considered to be a highly significant diagnostic indicator of this condition. The oral significance of HIV infections is discussed later in this chapter.

5. *Trauma.* This particularly involves the mucosa lying below ill-fitting or old dentures and is by far the most frequent cause of oral candidiasis.

The clinical presentation of oral candidiasis covers a wide range and the clinical conditions are perhaps best considered under the headings adopted by Lehner. These are as follows.

(1) Acute pseudomembranous candidiasis (thrush).

(2) Acute atrophic candidiasis.

(3) Chronic atrophic candidiasis (denture sore mouth).

(4) Chronic hyperplastic candidiasis (*Candida leukoplakia*). In this condition there is not only evident candidal invasion of the oral epithelium, but also marked hyperplasia of the epithelium itself, often with the production of epithelial atypia. There is considerable discussion as to whether this lesion represents a secondary infection of a pre-existing epithelial abnormality or an epithelial reaction to candidal invasion. However, from a clinical point of view there is no doubt that this lesion must be considered amongst those generally regarded as potentially premalignant and, for this reason, it will be fully discussed in Chapter 5.

The variation in host response to *Candida* infection may show in the nature of the lesions produced. If, for instance, the host resistance is moderately high a localized lesion will be produced, but if, on the other hand, the host resistance is low then the resulting lesion is likely to be more diffuse.

## Acute pseudomembranous candidiasis (thrush)

Except in very young babies, in whom the full immunological defences have not become active, the onset of this condition is practically always a sign of a debilitating generalized condition. The pseudomembrane consists of a network of candidal hyphae containing desquamated cells, organisms, fibrin, inflammatory cells, and debris. This lies on the surface of the tissue and candidal hyphae penetrate superficially into the epithelium to provide anchorage. Clinically, this appears as a thick white coating or series of patches on the affected tissue (Fig. 3.4). The pseudomembrane can be wiped away and, since the more superficial layers of the epithelium may be included, a red and bleeding base is left behind. This may be considered a reasonable preliminary clinical test to distinguish thrush from other white lesions of the mucosa in case of doubt and confirmation may be obtained by taking a direct smear from a scraping of the lesion. This may be fixed by gentle heat and at once stained, using the periodic acid–Schiff reagent, the hyphae are readily distinguished under the microscope.

Any of the mucosal surfaces of the mouth may be affected by thrush, as may the posterior pharyngeal wall. In this last instance the condition must be taken particularly seriously, since extension into the oesophagus and trachea is possible and may prove fatal. However, this is likely to occur only in the severely debilitated patient, although the pre-existing condition may not have been previously recognized. An observed case of rapidly spreading thrush following tooth extraction in a patient with unsuspected aplastic anaemia led to a fatal conclusion as a result of widespread candidaemia, fortunately a rare combination of circumstances, but one which illustrates the care which must be taken in dealing with this form of candidiasis. In a few patients laryngeal candidiasis of a less active form may be associated with oral lesions; this has been noted in patients with antibiotic stomatitis and in those taking oral steroid preparations. In general, the laryngeal discomfort and hoarseness which this produces clears with treatment of the oral lesions.

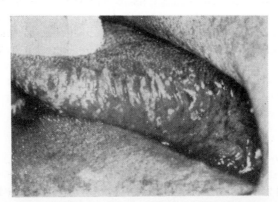

**Fig. 3.4.** Acute pseudomembrane candidiasis (thrush) on the tongue.

Treatment of thrush may be carried out by the use of nystatin or amphotericin B creams, ointments, lozenges, or suspensions. Any of these, used three or four times daily, will deal with the infection in the great majority of cases. It is evident, however, that treatment of symptoms alone is insufficient and that steps must be taken to determine any predisposing cause. In some patients the essential underlying factor is understood and under reasonable control, but in others, in particular those patients with AIDS, the debilitating nature of the condition may be expressed in virtually continuous candidiasis of the mouth. In these circumstances it is necessary to maintain long-term antifungal treatment; at the present time there seem to be no cases reported in which the organisms become antibiotic resistant. A particular problem arises in the case of patients with leukaemia or other neoplastic diseases under treatment with cytotoxic drugs and steroids, in these cases not only the underlying disease process, but also the therapy, tend to predispose towards candidiasis.

The polyene antibiotics, nystatin and amphotericin B, are well established and relatively free from side effects when used locally. As has been pointed out above, resistance is effectively unknown. However, the systemic use of amphotericin by the intravenous route (the only available systemic formulation of the polyenes) must be approached with great caution because of a wide range of possible serious side-effects including renal damage, hypersentivity reactions, and thrombophlebitis at the site of injection. The newer imidazole antifungal agents have not replaced the polyenes for general use, but have very useful properties. The locally acting substance miconazole is available as an oral gell, lozenges, and cream. Apart from its antifungal activity it has a limited antibacterial effect and it is suggested that it exerts a localized immune stimulatory function which enhances its antimicrobial action. Ketaconazole is a systemically acting imidazole which may be used in more generalized candidal infections. It is far safer than intravenous amphotericin B, but none the less, has a marked hepatotoxic effect in a few patients. If the decision is made to use systemic therapy with this drug, careful monitoring of liver function must be carried out. This form of therapy is contraindicated for the vast majority of patients with oral candidosis. Apart from the risk factor, ketoconazole binds onto salivary protein and is less effective in many cases than local preparations of miconazole. However, trials are now taking place with newer imidazoles which are less toxic than ketoconazole and may prove to be the choice of therapy when long-term systemic use is indicated. There are some reported cases of candidal resistance to the imidazoles and some of hypersensitivity reactions.

## Acute atrophic candidiasis

As has been pointed out by Lehner, this resembles thrush without the overlying pseudomembrane. As would be expected from this description the epithelium is thin and atrophic with candidal hyphae embedded superficially in the epithelium. The condition may follow or be concurrent with thrush, or may occur as the only manifestation of the infection. Clinically, the mucosa involved is red and painful; this is, in fact, the only variant of oral candidiasis in which pain and discomfort is marked. This form of candidiasis is often seen in patients undergoing either prolonged antibiotic or steroid therapy, and in this second group the oral discomfort may present a great problem. It is rarely possible for those patients on steroids to discontinue treatment and the continuous lowering of the resistance of the tissues to infection makes the treatment by the usual means relatively ineffective. However, there is no alternative to the carefully judged use of locally acting antifungal agents as the only available method of symptomatic treatment for these patients.

The 'antibiotic sore tongue' is a special case of acute atrophic candidiasis. As mentioned above this is a painful condition which responds relatively slowly to treatment. In a patient known to be susceptible to this condition and for whom it is necessary to prescribe further antibiotic treatment, it would be wise to add oral antifungal treatment to the regime.

## Chronic atrophic candidiasis (denture sore mouth)

This is by far the most common form of oral candidiasis. It has been found in 24 per cent of randomly selected denture wearers and represents the end result of secondary candidal infection of tissues traumatized by the continual wearing of a dental appliance. This need not be ill fitting, although most of the dentures involved are old, and, indeed, the appliance need not of necessity be a denture; an orthodontic plate may produce a similar result. The clinical picture is of a marked redness of the palatal mucosa covered by the appliance (the equivalent mandibular denture bearing area does not become involved) often with a sharply defined edge (Fig. 3.5). If a relief area is present on the denture there may be a corresponding area of spongy granular looking tissue (Fig. 3.6), but

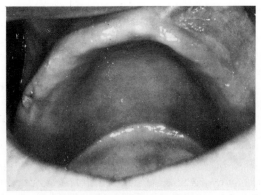

**Fig. 3.5.** Denture sore mouth. The sharply delineated nature of the affected palatal tissue is evident.

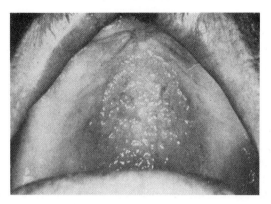

**Fig. 3.6.** Denture sore mouth, with a proliferative reaction below a relief chamber.

otherwise the affected mucosa is smooth. There is rarely any complaint of soreness by the patient in spite of the angry appearance of the tissues. It has been quite clearly demonstrated that denture sore mouth affects women much more frequently than men, this representing a true difference in incidence, not only in those reporting for treatment. Although endocrine influences have been suggested as the reason for this no fully convincing explanation has as yet been given. Because of its restricted distribution to the denture-bearing area denture sore mouth is still, from time to time, mistaken for an allergic reaction to acrylic resin, a condition which is, in fact, so rare as to be almost negligible.

By taking swabs or carrying out direct inoculation from the fitting surface of the denture involved it is practically always possible to isolate a heavy growth of candida. Swabs from the mucosal surface may also provide a prolific growth, but biopsy shows few candidal hyphae within the epithelium. In spite of this, however, significantly raised serum antibodies to candida have been found in the serum and saliva of patients studied, this implying a rather more active form of tissue involvement than would occur simply by growth of *Candida* in the space between the denture and the tissues.

Since there are two contributing factors to this condition (tissue trauma and infection) treatment must be directed at both in order to obtain rapid resolution of symptoms. However, the more important part of the treatment is undoubtedly the elimination of trauma by the adoption of suitable prosthetic techniques, the use of tissue conditioners being particularly valuable as an initial treatment. The use of an antifungal cream on the fitting surface of the appliance may prove a useful adjunct to treatment and helps to speed the resolution of the abnormal tissues. Careful and regular cleaning of all surfaces of the denture is also of great importance.

## Candidal angular cheilitis

A high proportion, but not all, of lesions of angular cheilitis are infected by candida, others are infected by bacteria. Candidal angular cheilitis is closely associated with denture sore mouth—it occurs in 80 per cent of such patients—and usually occurs in patients with deep folds at the angles of the mouth, often the result of faulty denture construction or of long-term alveolar changes below old dentures. However, a naturally deep angular fissure does occasionally occur in dentate patients and they may also develop angular cheilitis, sometimes intermittently as a response to some minor generalized condition such as a cold.

Clinically, the condition presents as an area of inflamed and cracked skin at the angles of the mouth (Fig. 3.7). There is often crusting and movement of the mouth may cause further cracking and bleeding. The affected tissue is heavily infected by *Candida* which are presumably able to thrive in the abnormal conditions of humidity, conditions brought about by the leakage and retention of saliva in the area. In long-standing cases granulomas may form adjacent to the folds and maintain the conditions of deep fissuring permanently. In order to eliminate these fissures it is sometimes necessary, as part of a complete treatment plan, to carry out a surgical removal of the granulomatous lesions, but in the vast majority of cases satisfactory results can be obtained by eliminating the folding of the skin by simple mechanical means. In the case of edentulous patients this is attained by careful adjustment of the vertical dimension or by adjustment of the contour of the dentures, whilst in dentate patients a removable appliance may be constructed to serve the same purpose. If this mechanical aspect of the treatment is combined with the use of an antifungal cream the combined effect is usually sufficient to eliminate symptoms. It has been shown that, in patients with denture sore mouth, treatment of this with antifungal antibiotics also frequently leads to a resolution of an associated

Fig. 3.7. Chronic angular cheilitis with candidal infection.

angular cheilitis. Presumably, this is due both to an overspill of the antibiotic via the saliva and also the elimination of the palatal reservoir of infection.

Evidently, if the organism involved in a lesion of angular cheilitis is not *Candida* then therapy must be modified. In the great majority of cases where the organism is not fungal it is likely to be *Staphylococcus aureus*; in this case fusidic acid cream is an effective agent. In some instances both *Candida* and staphylococci are involved; in these circumstances miconazole cream with its combined antifungal and antibacterial activity is particularly useful.

The subject of angular cheilitis will be referred to in later chapters since there are some instances in which the aetiology is different from that described above. In other cases, the appearance of angular cheilitis may mark the onset of significant generalized disease. For instance, in haematological diseases such as pernicious anaemia or in diabetes mellitus, otherwise unexplained angular cheilitis may be an early indication of the debilitating effect of the systemic condition as it may also be in AIDS.

## Other chronic candidal infections involving the mouth

Apart from the more common forms of oral candidiasis there are a few cases in which the mouth may be involved in more extensive chronic infections. These conditions may be classified as follows:

### Chronic mucocutaneous candidiasis

Apart from the more common forms of oral candidiasis there is a spectrum of conditions in which candidiasis of the oral cavity, the skin, and other structures such as the fingernails may occur, with or without association with other generalized disease processes. This group of conditions is generally known as chronic mucocutaneous candidiasis (CMC). The oral lesions in CMC are initially thrush like, but eventually abnormalities of keratinization of the oral epithelium may occur. The skin lesions may include widespread and disfiguring lesions of the face and scalp. Granulomatous lesions of the lips may also occur, similar to those which affect the skin. However, the range of structures involved is widely variable and a form essentially affecting the oral cavity and the fingernails only, without other systemic disease implications, is the most common. A variant (familial chronic mucocutaneous candidiasis) is genetically determined and is transmitted as an autosomal recessive condition. A classification was proposed by Wells.

Group 1. Familial CMC with genetic linkage as mentioned above.

Group 2. CMC presenting with widespread lesions over the body. There is a tendency to develop other infections although a primary immunodeficiency is hard to demonstrate. This form does not have an obvious genetic basis.

Group 3. CMC associated with endocrine disorders (endocrine candidiasis syndrome). This form is often inherited on an autosomal recessive basis.

Group 4. Late onset CMC which is usually severe and widespread, without a
genetic basis and in male patients almost exclusively.

Other classifications of CMC may be used and, indeed, the subject is not very
well understood.

Treatment of the various forms of chronic mucocutaneous candidiasis currently depends on the use of imidazoles. In spite of the possibility of liver damage,
the long-term use of ketoconazole has completely changed the outlook for many
patients with widespread disease. In those patients with predominantly oral
lesions the use of local miconazole preparations may be very effective. In these
circumstances the imidazoles are much more effective than the polyene antifungals unless resistant strains develop.

### Chronic diffuse (late onset) candidiasis

This occurs in older patients, the lesions consisting of widespread superficial
crusting areas over the limbs, trunk, and scalp. These patients, who are also
liable to develop recurrent chest infections, cannot be shown to have a primary
immunodeficiency.

### Endocrine candidiasis syndrome

In this syndrome white candidal plaques in the mouth and candidal infections of
the nails are associated with disorders of the parathyroids or adrenals. The
mechanism of this association is not understood, but the whole syndrome is
inherited as an autosomal recessive trait, and thus differs from the situation in
which candidiasis is associated with non-hereditary endocrine abnormalities.

A few other rare manifestations of candidal infection occur in and around the
mouth. Two of these, cheilo-candidosis and midline glossitis, are described in
Chapter 4.

## Viral infections

A number of viral infections may occur in or around the mouth. These do not
have very clearly defined clinical characteristics and it may not always be easy
to make a precise diagnosis. However, virtually all of these infections present as
vesicular lesions of the oral mucosa, the vesicles showing a marked tendency to
break down with the production of ulcers and erosions.

Laboratory confirmation of the exact nature of the infection is difficult and
slow. There are three methods which may be employed in diagnosis: examination of the cytology of cells scraped from the lesions, antibody measurements,
and direct virus culture.

Exfoliative cytology is often successful in demonstrating the essentially viral
origin of the lesions by the appearance in the smears of epithelial cells with

swollen distorted nuclei containing viral inclusion bodies. These are not specific to any particular virus. Measurements of specific antibodies may be carried out, not as an absolute determination but by measuring the rise in antibody titre over a period of 10–14 days after the onset of symptoms. Clearly, this is of no help so far as initial treatment is concerned. The third diagnostic technique—growth of the virus for identification—is also slow and complex. At the present time it is true to say that the initial diagnosis of oral viral infections must, of necessity, be clinical in the majority of cases.

It should perhaps be added that, in a very few centres, electron microscopy may be used to identify specific viral infections. This cannot be considered as a routine method of diagnosis at the present time. Similarly, immunofluorescent diagnostic techniques are available only in a very few centres.

Specifc treatment for viral stomatitis, as for other viral conditions, is as yet not easily available and symptomatic treatment is generally used. This, however, (in the form of a 2 per cent tetracycline mouthwash) is often remarkably effective in reducing the acute symptoms and giving comfort to the patient by elimination of secondary infection. The essential problems in treatment of viral lesions are two in number. The first is that no substance has, as yet, been produced which is entirely active against any virus when applied *in vivo*. The natural defence at cellular level is a substance (interferon) which, when synthesized by the cells at risk, prevents the increase of the virus particles by interfering with the essential molecular replication process. Human leukocyte derived Interferon can be used to treat life threatening viral infections in a very restricted number of patients—for instance, massive infections in highly immunosuppressed patients—but this is far from a routine procedure. The two substances currently most widely used in the specific treatment of viral infections in and around the oral cavity are idoxuridine and acyclovir. However, the field is one in which there is intense research activity and in which there may be significant change in the near future. Idoxuridine is an analogue of thymidine and acts by interfering with the intracellular replication processes of the virus. It is available for local use only; it is too toxic a substance for systemic use. Acyclovir, on the other hand can be used systemically, either by the oral route or, if needs be, intravenously, as well as by local application. The antimetabolic activity of acyclovir appears to be restricted to the cells containing viral components. The second difficulty in the treatment of many viral lesions is that replication of the virus takes place, with consequent cell damage, before symptoms appear. Thus, whatever treatment is applied, the essential lesion may have been irreversibly produced. It is for this reason that at the present time, much of the treatment of viral infections is of necessity aimed at reduction of symptoms rather than at the virus itself.

Idoxuridine is available in dilute aqueous solutions (0.5 per cent) but these are relatively ineffective in dealing with oral and perioral lesions, being initially designed for ophthalmic use. For use on skin a 5 per cent (or even 40 per cent) solution in dimethylsulphoxide (DMSO) is effective in suppressing viral replication and hence, if applied early enough, in aborting lesions. Acyclovir is available as a 5 per cent cream and as tablets of 200 mg as well as in the intravenous form mentioned above. Current practice in the treatment of *herpes simplex* and *zoster* is to use a combination of local and systemic therapy, both used frequently (five daily doses and applications).

Mention should be made of the antiviral agent AZT (Zidovudine) which has been introduced and is being much used as the only therapeutic agent available to affect in any way the progress of viral replication in HIV infections. As yet it has no other uses.

## Herpes simplex

The *herpes simplex* virus is a DNA virus of which two main groups are known in man, together with a number of slightly modified transitional types. *Herpes hominis*, Type 1, affects the oral mucosa, pharynx, and skin whilst *herpes hominis*, Type 2, predominantly involves the genitalia. The lesions produced by these two types of the virus appear to be identical, but there are suggestions that the long-term consequences of infection may be different since the Type 2 virus has been implicated in the production of cervical carcinoma. However, in the consideration of oral and facial lesions it is the Type 1 virus which is implied in the majority of cases (about 90 per cent in most centres). With the widespread recognition of the existance of genital herpes, patients now frequently ask about the type of virus involved because of anxiety about the route of infection. This information is not usually available since typing is a complex process and not routinely carried out. Additionally, the existence of 10 per cent type 2 infections in the oral cavity and an equivalent number of type 1 infections in the genital area makes the determination of the route of infection no more than speculative.

Primary infection frequently occurs via the oral mucosa and lips with the production of a primary herpetic stomatitis, although it is likely that many primary infections are relatively insignificant and, therefore, go unrecognized. A fully developed herpetic stomatitis affects patients in two main age groups, young children and young adults. An acute herpetic stomatitis represents the primary infection and, although subsequent immunity may not be complete, it is only very rarely that a second acute attack follows. If, in fact, a second episode of acute herpetic stomatitis apparently occurs, the diagnosis should be critically reviewed and a differential diagnosis including erythema multiforme considered.

The patient with acute herpetic stomatitis often gives a history of recent exposure to a patient with a herpetic lesion; the incubation period is about a week. The initial symptoms are of malaise with tiredness, generalized muscle aches, and, sometimes, a sore throat. At this early stage the submandibular lymph nodes are often enlarged and tender. This prodromal phase may be expected to last for a day or two and is followed by the appearance of oral and, sometimes, circumoral lesions. Groups of vesicles form on the oral mucosa and rapidly break down to produce shallow ulcers. Although the vesicles may be relatively small, the breakdown of confluent groups may result in the formation of large areas of ulceration (Fig. 3.8). The distribution of the lesions is variable; any of the oral mucosal surfaces may be involved, and from time to time unilateral lesions or lesions restricted to a relatively small area may occur. It has been customary in the past to ascribe such restricted lesions to *herpes zoster* rather than to *herpes simplex* infections, but serologically established restricted lesions caused by *herpes simplex* are not rare. Apart from the ulcerated areas the whole of the oral mucosa may be bright-red and sore and, particularly in young children, there may be a marked gingivitis which closely resembles that of acute leukaemia. It is evident that such a child should be screened for haematological changes. Apart from the intra-oral lesions there may be lesions of the lips and circumoral skin, which, because of the relative stability of the skin compared to the oral mucosa, may retain a much more obviously vesicular appearance. In a few cases the primary infection may become widespread and disseminated throughout the body, so that encephalitis, meningitis, and other life-endangering conditions may follow. Such potentially fatal cases have been mainly reported in patients with predisposing conditions; for instance, there is a high degree of association between disseminated herpes and kwashiorkor in African children. The possible relationship between viral infections and the initiation of cancrum oris in debilitated children has been mentioned previously in this chapter.

In the vast majority of untreated cases there is a slow recovery from the symptoms over a period of some 10 days, during which virus neutralizing

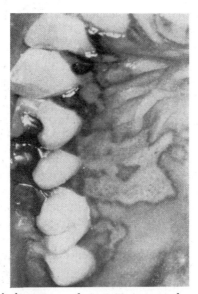

**Fig. 3.8.** Palatal ulceration and gingivitis in acute herpetic stomatitis.

antibodies appear in the blood. These are at their maximum at about 14–21 days after the onset of symptoms, following which they fall to a lower level which in most cases, is maintained throughout life. The role of the cell-mediated immune system is not clear. There is evidence that functional T-lymphocytes are necessary to clear the virus after a recurrent infection has occurred. However, it seems that there is a poor response of immunocompetent cells to the area of infection in the earlier stages of the lesions. The clinical recovery of the patient from the acute phase does not correspond to complete elimination of the virus, which remains within the cells of the host in a latent form. There is a great deal of doubt as to the exact location of this latent virus since intact virions are not detectable within the tissues, but it seems probable that only viral DNA remains within the host cells from which virions may be reactivitated even in the presence of humoral antibodies. The most significant site for the latent virus is thought to be in the sensory nerves serving the affected area; in the case of oral and perioral lesions, in the trigeminal nerve. The exact site is not known, but it is possible that it is the trigeminal ganglion itself.

The treatment of acute herpetic stomatitis depends almost entirely upon the elimination of secondary infection by the regular use of 2 per cent tetracycline mouthwashes which, as mentioned above, are remarkably effective in reducing the symptoms. Specific antiviral treatment has not been widely reported and, in view of the relative success of non-specific antibiotic treatment, it would seem to be contra-indicated in most cases. However, in a few cases (such as patients on immunosuppressive therapy), the use of systemic Acyclovir is indicated to minimize the possibility of generalized dissemination of the infection. It may well be the case that, in the future, such therapy may become standard for all patients. As yet little is known about the long-term benefits of acyclovir therapy.

## Recurrent herpes

Following the resolution of the primary herpetic infection there is an approximately 50 per cent likelihood that recurrent lesions will follow, regardless of the intensity of the primary attack. The usual site of the recurrent lesions is on or near the lips, although less commonly the skin and mucosa of the nose and nasal passages are involved, as may be, occasionally, almost any site on the face. In a given patient, however, the areas involved tend to remain the same in successive episodes. The recurrences may be provoked by a wide range of stimulae including sunlight, mechanical trauma and, particularly, mild febrile conditions such as the common cold. Emotional factors also play a part in precipitating recurrences in many patients, and it has been suggested that in these circumstances an excessive production of adrenaline may be involved. This speculation has received some confirmation from animal experiments and it is of interest to compare this theory with that propounded to explain the recurrence of AUG in tobacco smokers, previously referred to in this chapter.

In the prodromal stages of a recurrence the patient may feel a mild degree of tiredness and malaise. This is quickly followed by a period of irritation and itching over the area of the recurrence and within a few hours, vesicles appear surrounded by a mildly erythematous area. In a short time the vesicles burst and a scab is formed (Fig. 3.9). From this point the process is one of slow healing over a period of some 10 days, but secondary infection may occasionally delay the healing process and lead to the production of small pustules in the area. Healing is without scarring and the affected area returns, after a short period of erythema, to an apparently normal stage until the onset of the next recurrence. This may be after a period of a few weeks or even days in some individuals, but generally the intervals between recurrences are of the order of months, this evidently depending to a large extent on the degree of exposure to the particular stimulus involved.

The immunological factors associated with recurrent herpetic infections have been widely investigated and it has been well established that recurrence may occur in the presence of high levels of humoral antibodies which appear to exert no protective function. Although virus stimulated lymphocyte transformation is normal in affected patients, it has been shown that both the macrophage inhibition and lymphocyte toxicity are deficient. It has, therefore, been suggested that susceptibility to recurrent herpes may well be due to this deficiency in the cell-mediated immune mechanism. Among a very wide range of factors which contribute to the total resistance to infection, it has been suggested that there is a series of effector cells dependent on the presence of antibodies which play a special part in the immune defences against viruses. As yet, little is known about their mode of action.

Treatment of recurrent herpes is unsatisfactory. There is no therapeutic measure available which provides reasonably consistent results and which entirely prevents further recurrences. The application of 5 per cent idoxyuridine

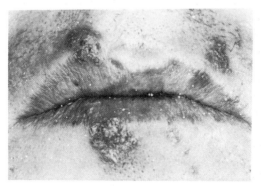

Fig. 3.9. Recurrent facial herpes ('cold sores').

in DMSO, which gives considerable relief from the irritation associated with the lesions if applied early enough and frequently enough (at least four times daily). It is reported that the average duration of the lesions is reduced from 9.5 to 3.5 days by this treatment. It is also suggested that, following the treatment, recurrence rarely takes place in the same site, although the present author's experience does not entirely confirm this claim. Acyclovir cream, used as early as possible and at least five times a day in the early stages, may also abort the lesions. This is now perhaps the most effective and acceptable form of treatment for recurrent herpes, although there is no real suggestion that a permanent cure may be obtained. At the present time trials are taking place to determine the comparative effects, on a short- and long-term basis, of local and systemic Acyclovir therapy. Perhaps the most useful preventive measure for those herpetic subjects susceptible to sunlight is the careful use of suitable screening creams and lotions if exposure to the sun is expected.

## Herpes zoster

The virus responsible for *herpes zoster* (herpes virus *varicellae*) is a DNA virus morphologically similar to the *herpes simplex* virus and apparently responsible for two completely dissimilar diseases in humans—chicken-pox and *herpes zoster*. There is, however, little evidence that contact with one of these diseases is responsible for the initiation of the other. It seems more likely that the *zoster* eruption represents the reactivation of the virus in a previously infected patient with only partial immunity, a situation parallel to that in recurrent *herpes simplex*. Much as in the case of *herpex simplex*, the *zoster* virus is thought to remain latent in the relevant sensory ganglion and to pass down the nerve to the skin or mucous membrane on reactivation.

Most patients with *herpes zoster* are middle aged or older (70 per cent over the age of 50), but it can occur in much younger patients and a few neonatal cases have been described. It is suggested that there is a greater incidence in male patients. Predisposing factors which have been suggested include a wide range of debilitating diseases and immune incompetence—either of iatrogenic or naturally occurring origin.

The characteristic superficial lesion of *herpes zoster* is a vesicular eruption in an area of distribution of a sensory nerve. The band-like distribution of shingles on the trunk is well known, having given the name to the virus (*zoster* = girdle). When the eruption affects the trigeminal nerve the facial skin and oral mucosa in the sensory area may be affected. Of the three divisions of the trigeminal nerve the ophthalmic is the most frequently involved, but the other two divisions are also not uncommonly affected. Occasionally, when the disorder affects one division of the trigeminal nerve the adjacent division becomes involved as well. Cervical nerves may also be affected. The initial symptoms are of pain and tenderness in the affected area—a discomfort much more severe than that experienced in *herpes simplex*. The prodromal phase may last for 2 or 3 days and is succeeded by the appearance of vesicles in a rash, which may be either sparse or so dense as to be almost confluent. Frequently, the vesicles appear over a period of days rather than together, and there may be marked secondary infection. When the ophthalmic division of the trigeminal nerve is involved there may be corneal ulceration which requires most careful management to avoid permanent scar formation. Within the mouth, the vesicles behave much as those of *herpes simplex* infections, rapidly breaking down to form ulcers. The distribution of the lesions intra-orally, unilateral and often confined to the area of a single branch of the trigeminal nerve, may give a clue as to the nature of the infection, but this cannot be relied upon. It has been pointed out earlier that *herpes simplex* infections (and, indeed, those of other viruses) may present with a similar restricted distribution. If untreated the vesicles and oral ulceration fade over a variable period from 2 to 4 weeks, or even longer. The skin vesicles may form firm crusts and, particularly if these are disturbed, marked scarring may occur. Following the fading of the rash the major complication of the condition—post-herpetic neuralgia—may appear. In this condition anaesthesia, paraesthesia, and trigeminal neuralgia-like pains may affect the area and may persist for a period of years, occasionally reappearing after a prolonged absence. This is a highly refractory condition which often fails to respond to any form of medical treatment and for which neurosurgery is occasionally advocated. The essential lesion is an irreversible fibrotic change in the neural bundle (see Chapter 11). In a very few patients the facial nerve may become involved during an episode of *zoster* reactivation, probably via the geniculate ganglion. Facial weakness, loss of taste sensation, and symptoms such as dizziness resulting from labyrinthine disturbance may occur. This is the Ramsay–Hunt syndrome. Vesicular lesions in this condition are most often seen on the palate and around the external auditory meatus. On the whole this is a self limiting condition which resolves with restoration of function, but in some patients there may be permanent facial weakness.

Treatment of the skin lesions is at present carried out with 40 per cent idoxyuridine in DMSO, with care being taken to avoid the eyes. Alternatively, Acyclovir cream, together with systemic Acyclovir (oral), may be used. In the case of the oral lesions, the same treatment as for *herpes simplex* stomatitis (2 per cent tetracycline mouthwash) is used, with the same rationale and with similar, if slower, results. Just as in the case of *herpes simplex*, the use of steroids in *herpes zoster* is controversial. It has been suggested that systemic steroids should be used to reduce the possibility of permanent nerve damage and, hence, post-herpetic neuralgia. However, the risks of this form of treatment, particularly in the already possibly immunocompromised patient, are evident. In a very few centres leukocyte derived interferon or *zoster*-immune globulin may be available to treat high risk patients. Perhaps one of the most important functions of the dental surgeon faced with a case of *herpes zoster* is to be aware of the possibility of corneal involvement and to place the patient under appropriate care if any symptoms, however mild, appear.

## Coxsackie virus infections

The Coxsackie group of viruses consists of a number of types, many of which are capable of producing significant infections in humans. Of these, two affect the oral cavity, although neither of the conditions can be considered at all common.

Herpangina, caused by Coxsackie A, Type 4 virus, is a mild infection seen predominantly (but not entirely) in children, which tends to occur in minor epidemics. The patient complains of a moderate degree of malaise and of a sore throat, while occasionally, there is also a minor degree of muscle weakness and pain. Small vesicular lesions appear in the posterior part of the mouth, in particular, on the soft palate. These lesions are not particularly characteristic of the condition, but resemble other virus-induced lesions and are recognizable only by their typical distribution. However, as has been pointed out, a viral stomatitis cannot be fully characterized in this way and it is only by serum antibody titres that a definite diagnosis can be made. This is a self-limiting condition, the lesions fade after some 3–5 days and complications are extremely rare.

Hand-foot-and-mouth disease (not the same as foot-and-mouth disease) is caused by another Coxsackie A virus, in this case usually Type 16 (but very rarely Types 5 and 10). A number of minor epidemics of this condition have occurred and been described in detail, one involving students and staff at a dental hospital. The oral lesions in this outbreak consisted of small ulcers, resembling minor aphthous ulcers, relatively few in number and distributed over the oral mucosa. However, oral lesions observed by the author, also in dental hospital personnel and confirmed by a full range of investigations, have in fact consisted of bullae which later ruptured to produce transiently painful erosions of the mucosa (Fig. 3.10). The lesions of the hands and feet consist of a red macular rash, each macule apparently surrounding a deep-seated vesicle. The constitutional symptoms experienced are of a mild nature: a slight malaise and minimal pain and discomfort. The symptoms resolve in about a week.

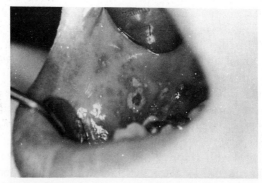

**Fig. 3.10.** Hand-foot-and-mouth disease, ruptured bullous lesion of buccal mucosa.

A few more severe infections by Coxsackie viruses have been reported, but in the majority of cases, the oral discomfort produced in these conditions is minor, and by no means as marked, for instance, as that in many cases of primary herpetic stomatitis. Treatment does not appear to be necessary since the disease is self-limiting.

## Infective mononucleosis (glandular fever)

Glandular fever is a quite common viral infection, particularly among young adults. The virus involved is the Epstein–Barr virus which, strangely enough, seems also to be associated with the formation of malignant lymphomas in Africa (the Burkitt tumour). The patient with glandular fever feels ill, has a fever and headaches, and enlargement of the lymph nodes which may extend over the whole of the body. The severity of these symptoms is very widely variable, ranging from a minor attack which may pass almost unnoticed to a condition requiring hospitalization. During the acute phases of the infection patients produce abnormal leucocytes in the peripheral blood which are diagnostic of the condition. There is also an immunological reaction (the Paul–Bunnel test) which depends on the fact that the serum from these patients agglutinates sheep erythrocytes.

In the early stages of this infection a sore throat and oral ulceration may be very troublesome. The ulceration is quite non-specific, but may be widespread (Fig. 3.11). As in similar circumstances symptomatic treatment of the oral ulcers with antibiotic mouthwashes is the only helpful procedure.

## AIDS

The acquired immune deficiency syndrome (AIDS) is one possible expression of infection by a RNA retrovirus which is thought to be new to man, the first clinical manifestations being recognized in 1981. The virus involved is fully

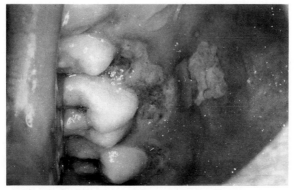

**Fig. 3.11.** Palatal ulceration in infective mononucleosis (glandular fever).

described as the human T-cell lymphotrophic virus III (HTLV III) although it is often referred to as the AIDS virus and, more correctly, as the human immunodeficiency virus (HIV). With the possible identification of other types of the virus the term HIV1 is coming into use (to differentiate between HIV2 and other possibilities). The viral infection is a slow process and it has been shown that in infected patients integrated provirions of DNA are present in the genetic material of some brain cells. Thus, these patients become the source of their own infecting RNA virions which are replicated in and then expressed from the cells concerned. The most significant effect of the virus is in the infection and consequent inactivation of a small, but highly significant subgroup of T-helper lymphocytes. The eventual result of this may be profound suppression of the immune response. It seems that this is by no means invariable and there are a considerable number of known carriers of the virus (and probably an even larger unknown number) who display no evidence of immunosuppression. It has been suggested that there may be a number of cofactors such as minor pre existing immune abnormalities or genetic susceptibility which may differentiate between the virus infected patients who do, or do not, develop the syndrome.

The country of origin of the virus is unknown although there is fairly strong presumptive evidence that the initial centre of activity was in central Africa. However, the problem is now a rapidly expanding one world wide although, quite clearly, prognostication in these circumstances is a very difficult process and there can be no degree of certainty about the numerous projections of future patient numbers which have been made. Transmission of the virus is currently thought to be predominantly via blood and blood products, and semen, Initially, it was thought that saliva was an important carrier of the virus, but it has recently been demonstrated that not only is the saliva of infected patients virus free, but that the saliva may contain an inhibitory substance. The earlier concentration on male homosexual behaviour as being by far the most likely method of transmission of the virus is now under question although this still seems to be a highly significant factor. Transmission by common syringe use in intravenous drug users is now thought to be even more important.

As yet there is no treatment available either for AIDS itself or for the related conditions apart from the inhibitor AZT mentioned earlier in this chapter. No other form of immunotherapy or antiviral therapy has been shown to have any effect on the progress of the condition and it is currently considered to be effectively untreatable.

There are two important factors in relation to the oral cavity of these patients. The first is concerned with the dental treatment of known virus carriers and the second with the recognition of possible infected patients. Numerous guidelines have been issued and are likely to continue being issued on the subject of the dental treatment of these patients. In general, these parallel the advice given for the treatment of known hepatitis B virus carriers. At the time of writing, however, no health care personnel have been identified as having been infected with HIV during the treatment of a known patient. This is in contrast to the situation in relation to the hepatitis B virus which has been shown to have been transmitted to numerous medical, dental, and other health workers from affected patients.

In the present context the most significant consideration is of the changes which may appear in the oral cavity of a patient with the syndrome which has not, as yet, been diagnosed. The likely oral changes are dependant on the increasing reduction in immune surveyance, presenting either as infections or as neoplasms. The most likely infection is candidiasis. As has been pointed out previously in this chapter, oral candidal infection should always be considered as an indicator or generalized ill health and, in the case of the early AIDS patient, it may be the earliest presenting sign. Other viral, fungal, and bacterial oral infections are much less likely, but do occur in some patients. The otherwise unexplained onset of such an infection, particularly when associated with other persistant signs and symptoms such as generalized lymph node enlargement, malaise, intermittent fevers, and weight loss should arouse suspicions. A leukopaenia or other unexplained abnormalities on routine blood examination should further strengthen these suspicions and lead to a full investigation by a competent authority. In these circumstances blood for further serological and other studies should be taken with the adoption of the precautions currently recommended.

The neoplasm which is most likely to occur in the AIDS patient is Kaposi's sarcoma (Fig. 3.12). The mouth and, in particular, the mucosa of the hard palate is a common site for this lesion which is, in fact, a form of diffuse lymphoma rather than a discrete neoplasm. The lesion is described as a pigmented, non-painful slightly nodular lesion of the mucosa with a characteristic histological appearance. Before the recognition of AIDS this was considered to be a rare lesion, confined to elderly patients of several restricted racial groups (Bantu, Jewish, and Italian) or to patients on immunosuppressive therapy. Its appearance in patients of other kinds is now said to be virtually pathognomic of

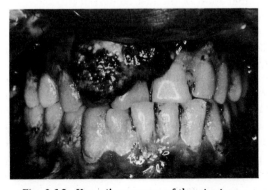

Fig. 3.12. Kaposi's sarcoma of the gingivae.

active AIDS. Other neoplastic lesions, such as squamous cell carcinoma and melanoma have also been described in AIDS patients, but there is no real evidence of other than a random association as yet. Oral leukoplakia, and in particular a characteristic type ('hairy leukoplakia') has also been described as a possible indicator of the condition.

The 'hairy leukoplakia' has been described in a large number of patients who are HIV antibody positive. It occurs on the lateral margins of the tongue and has a folded, corrugated or 'hairy' appearance. Although *Candida* may be associated with the lesion, it does not seem to be a '*Candida* leukoplakia' or chronic hyperplastic candidiasis as described earlier in this chapter. Strangely enough, the lesion appears to be associated with the Epstein–Barr virus as demonstrated both by electron microscopy and by immunological studies. It is suggested that hairy leukoplakia is a powerful clinical indicator of immunosuppression in HIV infected patients and that it is a predictor of the eventual onset of AIDS. This does not seem to be a lesion with significant premalignant potential; no cases of malignant transformation in hairy leukoplakia have as yet been described.

The gingival response to AIDS takes two forms. The first of these is persistant and recurrent acute ulcerative gingivitis. The second is a rapid degenerative periodontitis, the true nature of which has not as yet been discovered, but which seems to depend on the reduced ability of leukocytes to maintain periodontal stability.

The only test at present available for the presence of the HIV virus is an antigen–antibody reaction. Evidently, this test (which is currently being refined to a reasonable degree of specificity) does not act as a diagnostic indicator of AIDS itself, but of the presence of the virus and, hence, of the possibility of the eventual onset of the disease or of its transmission. In patients developing the virus related complex, laboratory tests may detect a decreased number and proportion of T-helper cells together with a wide range of changes in blood cell cytology and in indicators of immune competence.

It is quite evident that a condition such as AIDS and the other HIV-related conditions which have been only recognized and investigated during the few years since the publication of the last edition of this book cannot be described with any degree of confidence. Many highly significant changes in consensus opinion regarding various aspects of the problem have occurred in this period and research activity is intense. It may well be that in the future a more optimistic prognosis for these unfortunate patients may apply.

## Viruses and neoplasia

That there is a causal relationship between viruses and some animal tumours and leukaemias has been established for some time. As mentioned above, in the human field a strong association has been established between the Epstein–Barr virus (a herpes virus) and the Burkitt lymphoma. An almost equally strong association between the hepatitis B virus and liver cancer has been demonstrated. The causal association between papilloma viruses and human neoplasms, both benign and malignant is also generally accepted.

Many attempts have been made to prove that herpes viruses are involved as a cause of cervical carcinoma, but in fact, the only proof is of an association which, in view of the multifactorial nature of the problem, might not be causal. Similarly, there have been a number of attempts to associate oral carcinoma and potentially malignant conditions with viral infection, but as yet the evidence is largely conjectural and based on immunological studies which are capable of differing interpretations.

# 4

# Recurrent oral ulceration

MANY patients suffer from recurrent ulceration of the oral mucosa, ulceration which may prove painful and, in some individuals, incapacitating. Although in a few of these patients the ulcers may be the result of any one of a wide range of conditions (for example, certain skin diseases), the most common lesions are those of a well-defined group to which the term 'recurrent oral ulceration' (ROU) is often applied. Among the many synonyms adopted for this group of conditions 'recurrent aphthous stomatitis' (RAS) is perhaps the most popular. As many as one-fifth of the population may be affected in this way at some time. There are many clinical variants, but most cases of recurrent ulceration can be grouped into one or other of three broad classifications based primarily on the clinical features of the condition. These three groups are:

(1) minor aphthous ulcers;

(2) major aphthous ulcers;

(3) herpetiform ulcers.

Many synonyms exist for these terms, but since they were introduced by Truelove and Morris-Owen and used in a subsequent series of important papers by Lehner, this nomenclature has become generally accepted. In general, the ulceration is confined to the mouth, but in a few patients other mucous membranes may be involved. This situation and the reasons for this multiple involvement will be discussed later in this chapter. The diagnosis of ROU is not usually difficult and may be deduced, in most cases, from the history and the characteristic clinical appearance.

## Minor aphthous ulcers (MiAU)

This is the most common form of recurrent oral ulceration, some 80 per cent of patients have lesions of this type. It is reported that 56 per cent of the patients are females and that the peak age of onset of the ulcers is in the second decade (10–19 years). However, many patients experience their first lesions at an age well outside these limits; indeed, it is by no means uncommon to find this form of ulceration in much younger children. Similarly, some individuals initially experience the ulceration late in life, although it has been shown that 90 per cent of patients have had their first episode of ulceration before the age of 40 years. Even in those patients who have experienced the onset of ulceration in the second decade, there is no evidence of a correlation between the occurrence of the ulcers and the onset of puberty. Several attempts have been made to define the patients affected in terms of such factors as social class, intelligence quotient, and nature of employment, while other surveys have been carried out in an attempt to show psychiatric abnormalities, but no convincing positive results have been obtained.

In its most characteristic form MiAU presents the picture of a number of small ulcers (one to five) appearing on the buccal mucosa, the labial mucosa, the floor of the mouth, or the tongue. The palatal soft tissues are rarely involved, the exception being the area of the incisive papilla following an episode of mild trauma. Moreover, the ulcers are concentrated in the anterior part of the mouth; the pharynx and tonsillar fauces are rarely implicated in this form of ulceration. The prodromal stage of ulceration varies from patient to patient, but there is usually a sensation described as burning or pricking for a short period before the ulcers appear. Following this phase, ulceration occurs directly by loss of the epithelium; some cases have been described in which an initial vesicle has been seen, but this is quite exceptional. The ulcers are usually less than 1 cm in diameter and, in most instances, their size is about 4 or 5 mm. However, the classification of 'minor' ulceration does not depend on the dimensions of the lesions alone, but on the general picture; it is quite possible to have large minor ulcers and small major ones. The appearance of the ulcers is grey–yellow, often with a red and slightly raised margin and, unless influenced by the site (as in the depth of the buccal sulcus where they appear elongated) they are approximately round (Fig. 4.1). The ulcers are painful, particularly if the tongue is involved, and may make eating or speaking difficult. If the lips are implicated, there may be a minor degree of oedema in the surrounding area but this is not common. Lymph node enlargement is seen only as a response to secondary infection in severely affected patients and is not a routine finding. The course of these ulcers varies from a few days to a little over 2 weeks, but by far the most common

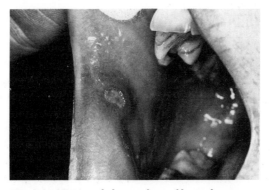

Fig. 4.1. Minor aphthous ulcer of buccal mucosa.

duration is of the order of 10 days. After this period the ulcerated areas re-epithelialize and heal over an interval of some days, but resolution may or may not be simultaneous in all the ulcers of a group. In MiAU healing occurs without scar formation; thus if scars should form, it is probable that the condition is not the minor form of ulceration, but the major type. Following healing of the ulcers, there is an ulcer-free interval which is widely variable—3–4 weeks is most common—but many patients are able to predict with some degree of accuracy the periodicity of the condition. In a few patients, however, the recurrence of the ulcers appears to be entirely random and in some cases there may not be an ulcer-free period between attacks, i.e. the ulcerative episodes overlap in time.

The nature of the possible precipitating factors in this condition is far from clear, but there is little doubt that the site of a specific ulcer or group of ulcers may be determined by some form of minor trauma. The occasional involvement of the incisive papilla has been mentioned; this occurs most often in patients in whom the arrangement of the dentition is such as to subject the area of the papilla to injury from the lower teeth. Minor toothbrush trauma may also determine the site, as may quite mild soft tissue damage from an irregular tooth or restoration. In such examples of precipitating trauma, however, it would seem that it is only the site which is selected by the effect of the injury since the basic aetiology of the ulceration is quite unrelated to trauma. However, in a patient subject to these ulcers a more acute form of trauma may precipitate an unusually large and painful ulcer, apparently combining the characteristics of a traumatic and an aphthous ulcer. Apart from physical injury, it is generally accepted that mental stress may be a precipitating factor in the occurrence of severe attacks; for instance, students under the strain of an examination are frequent sufferers. Generalized ill-health is also a significant factor in precipitating the ulceration, and it has recently become clear that a wide range of haematological abnormalities—insufficient to produce other recognizable symptoms—may be associated with oral ulceration of this kind. It follows, therefore, that a full haematological screening procedure should be considered for these patients. The relationship between MiAU (and other forms of recurrent ulceration) and coeliac disease is discussed in Chapter 8. There is an increasing body of evidence to show that, in some patients, otherwise asymptomatic coeliac disease may accompany recurrent oral ulceration and that, if the gluten enteropathy is treated by diet, the oral ulceration will cease. Since the simplest first diagnostic measure to detect the presence of coeliac disease is a haematology screen for possible malapsorption, the importance of this test is evident. The eventual importance of this association is yet to be fully assessed, but it would seem that a significant proportion of patients showing minor abnormalities on haematological screening are of this type. Possible reasons for the association are discussed in Chapter 9, but it must be acknowledged that the full interrelationship between the various factors is as yet far from obvious. In some, but by no means all, female patients the episodes of ulceration are accurately predictable in terms of the menstrual cycle, the usual onset of ulceration being 4 or 5 days before menstruation. In other female patients, but again not all, previously active ulceration may remit completely during pregnancy. Other well described phenomena are remarkably difficult to explain; for example, the fact that in some patients the onset of ulceration appears to follow cessation of a smoking habit and that the ulcers go into remission if the smoking habit is recommenced.

In general, the diagnosis of MiAU is by examination and clinical history; biopsy of the ulcers merely demonstrates a non-specific histology and contributes little to the actual diagnosis. However, in a few cases (particularly when there has been superimposed trauma), the ulcers may appear with ominously rolled edges and, should there not be the expected rapid regression, biopsy may be essential to exclude the possibility of malignancy. In the majority of cases the routine screening tests which may be carried out do not indicate any specific abnormality and, therefore, are of little help in confirming the clinical diagnosis.

The cause of aphthous ulceration is not known with certainty. Infections by viruses and by bacteria of various kinds have been suggested as the causal factor, but careful studies to determine a consistency present infective agent have been unproductive. In particular, the herpes virus has been implicated in the production of aphthae, but the evidence has proved entirely negative; the virus has never been consistently identified in the ulcers, whilst histological study of developing ulcers demonstrates none of the cellular changes characteristic of viral infections. Of the bacteria which have been implicated *Streptococcus sanguis* and its L form have been most often mentioned and it has been suggested that the ulceration may be due to a hypersensitivity reaction provoked by that organism. Immunological studies do not substantiate this.

However, there is evidence that abnormalities of the immune system are associated with the production of aphthous ulcers. Little evidence of allergy to external antigenic substances exists (although the relationship to gluten enteropathy is unexplained) whereas there is support to suggest that the mechanism causing the ulceration is autoimmune in nature. It has been demonstrated by Lehner that circulating humoral autoantibodies and sensitized T-lymphocytes active against the oral mucosa, circulate in patients with aphthous ulcers. It has also been shown that, although the humoral antibodies maintain a relatively constant level, the active T-lymphocytes undergo a phase of rapid proliferation just before the onset of ulceration. It has also been demonstrated that the lymphocytes from affected patients exert a toxic affect on oral epithelial cells, and there seems little doubt that at an early stage in the production of an ulcer, before tissue destruction occurs, there is a dense infiltration of lymphocytes at the site, preceded by intraepithelial deposition of immunoglobins. Thus, it would seem to be established that an autoimmune mechanism exists which would explain, at least to some extent, the production of aphthous ulcers. As has been previously mentioned, there is no firm evidence for the production of immunologically active complexes by external substances, and so it would appear that the activity of the immune process is more likely to be due to a direct sensitization of the competent lymphocytes by displaced epithelium within the lymphoid tissue.

It has been suggested that the resulting lack of tolerance to self antigens may be increased by the stimulating effect of local infections, such as those of periodontal disease, as a result of which all the immune functions (including the production of autoantibodies) are stimulated. One aspect of this possible autoimmune process remains quite unexplained, the nature of the triggering mechanism which initiates the rapid increase in sensitized lymphocytes before the ulceration occurs and the nature of the feedback mechanism involved in reversing this effect. Indeed, the position has recently become less well defined and it seems likely that, although both T-cell and antibody-mediated cytotoxicity against epithelium must play some part in the production of aphthous ulcers, a simple autoimmune process may not be held completely responsible. It has been pointed out that many of the classic features of autoimmune processes do not apply to ROU, in particular, its lack of association with other autoimmune diseases and the absence of a range of autoantibodies in the affected patients. The true nature of the inter-relationship between the autoimmune process and the gluten enteropathy leading to coeliac disease is also, as yet, unexplained.

The treatment of MiAU is an unsatisfactory procedure; the aetiology is, as yet, poorly understood and, consequently, the treatment is largely symptomatic. However, certain measures are helpful in controlling the extent of the ulceration, and these may give considerable relief by shortening the duration of the ulcers or by reducing the pain symptom. In the present state of knowledge there is no reasonable means of completely preventing the attacks, except in the case of patients with proven enteropathy. In some patients, however, it does appear that the successful abortion of an attack of ulceration, by any available means, may result in a longer period of remission than might be expected, possibly over several cycles of ulceration.

Approaches to the therapy of MiAU can be outlined as follows:

1. *The maintenance of strict oral hygiene.* This is of importance on two counts: In the first instance, it is obvious that the ulcers may be exacerbated and made more painful in the presence of local sepsis and resulting secondary infection. The second factor is the possibility of enhancement of the auto-immune process by the stimulating effect of local sepsis; such generalized stimulation of immune mechanisms by localized factors is well established in other situations. Insistence upon a satisfactory standard of oral hygiene may, in fact, prove difficult in the presence of painful ulceration, and patients may be reluctant to use a toothbrush which may precipitate ulceration by even a mild degree of trauma. Such an effect may also result from the presence of sharp-edged restoration of prominent natural cusps and so these irritants should be removed. Perhaps the most striking example of this form of irritation is the effect of a prominent maxillary canine in inducing repeated ulceration at the point of contact with the lower lip.

2. *The use of covering agents.* There are a number of available pastes and gels which can be used to coat the surface of the ulcers and to form a protective barrier against secondary infection and further mechanical irritation. Some

difficulty may be experienced in applying certain of these preparations, particularly to large ulcers and to those at the back of the mouth. It is also awkward to keep these 'coverings' in place on the tongue and lips where constant movement tends to wipe them off. In spite of these disadvantages, however, such simple remedies are often helpful in controlling the symptoms of MiAU. Some of these gels include agents such as choline salicylate which exert a non-specific anti-inflammatory effect, and so further reduce the pain and irritation associated with the ulcers.

3. *The use of antiseptics.* Antiseptic substances in a wide variety are available, suitably formulated as mouthwashes, pastilles, or lozenges, which may be helpful in temporarily reducing secondary infection. The response is variable, as might be expected in view of the wide range of organisms involved, but in many cases some degree of relief is achieved. Chlorhexidine mouthwashes are widely used for the symptomatic treatment of MiAU and are considered helpful by many patients although no clear formal demonstration of their efficiency has as yet been published.

4. *The use of topical antibiotics.* A much more effective measure in the relief of symptoms caused by secondary infection is the application of topical antibiotics. A mouthwash containing 2 per cent tetracycline or chlortetracycline is often highly effective in reducing the pain caused by severe ulceration and, as a result of the much less heavily colonized environment, the ulcers often heal more rapidly than otherwise. However, there are obvious disadvantages in the use of broad-spectrum antibiotics for this purpose, the risk of hypersensitivity reactions and the encouragement of growth of resistant organisms being the most important. Local secondary infection by opportunist invaders such as *Candida* seems to be much less of a problem, but as has been pointed out in Chapter 3, an 'antibiotic sore tongue' may very occasionally develop. In view of the obvious desirability of limiting the local use of antibiotics, this is a form of treatment which should be used sparingly and, in the majority of cases, as a single course of treatment for an isolated severe attack of ulceration.

5. *The use of topical steroids.* When properly used topical steroids are effective drugs in the treatment of MiAU. However, the patient response is variable and there are some individuals who gain little or no relief from their use. Steroids, used in this manner, have two modes of action. Their generalized anti-inflammatory action modifies, in a minor way, the progress of the ulceration at all stages and, to some extent, reduces the discomfort experienced. The second effect of steroids, i.e. the specific blocking effect of the T-lymphocyte (epithelial interaction), is much more important in the present context. Since the concentration of sensitized lymphocytes occurs before and during the early stages of ulceration, it follows that the drugs exert their maximum effect at this time. With the establishment of the ulceration and the fall in sensitized lymphocyte concentration, the specific blocking effect of the steroid becomes less important, and only the anti-inflammatory action of the drug operates. For this reason it is

important that the patient should understand that, whatever preparation is used, it is of maximum value at the time that the earliest prodromal signs of ulceration are noticed.

The drugs most commonly adopted for local oral application are hydrocortisone hemisuccinate (as pellets of 2.5 mg) and triamcinolone acetonide (in an adhesive paste containing 0.1 per cent of the steroid). Betamethasone sodium phosphate is no longer used in this way since a measurable degree of adrenal suppression may occur with relatively low dosages. This disadvantage does not apply to betamethasone-17-valerate and a regime of four applications daily of 100 µg has been maintained for long periods of time without any measurable adrenal suppression. However, suppression has been found to occur in some patients with dosages of 1.6 mg daily of betamethasone-17-valerate, 25 mg of hydrocortisone hemisuccinate, or 4 mg of triamcinolone acetonide. These are much larger than the usual clinical dosages in each case. Little is known about the long-term effect of small dosages of topical steroid therapy in children, although it does appear that the margin before adrenal suppression is reached is as great in children as in adults. However, it is evident that the use of steroids in children, even in small dosages, should be approached with caution. Side-effects from the use of local steroids in the mouth are not common, although the possible irritation of a latent peptic ulcer should be borne in mind. Infections are not often exacerbated by the use of hydrocortisone or triamcinolone, but there is a significant incidence of candidal infection (thrush) following the use of betamethasone-17-valerate. If this occurs, antifungal therapy must be instituted.

The use of systemic steroids is virtually never indicated in MiAU, but their occasional use in major aphthous ulceration will be considered below. Preliminary trials of small dosages of immunosuppressive drugs, other than steroids, have not been encouraging.

6. *Hormones.* In some female aphthous ulcer subjects, there is a complete remission of symptoms during pregnancy. Originally, it was hoped that the use of oral contraceptives would have a similar effect, but, in fact, the results seem to vary widely. In those patients in whom the ulceration appears to be precipitated by menstruation, treatment by relatively high dosages of oestrogens has been suggested as a method of control. However, in view of the concern at the incidence of thrombosis in patients on high-oestrogen contraceptives, the use of even higher doses to suppress oral ulceration would appear to be unwise.

7. *The use of topical anaesthetics.* Local anaesthetic lozenges may be used as a last resort to give a patient a brief period of relief from pain when, for instance, he or she is eating a meal. There are numerous preparations available in which a small dose of local anaesthetic (usually benzocaine) is combined with an antiseptic or a locally active antibiotic. These preparations, which are more often used for the treatment of sore throats, may find a limited use in the relief of discomfort due to MiAU.

8. A wide range of drugs has been used to treat MiAU with variable results. Among these, carbenoxolone and sodium cromoglycate have recently been most widely discussed. No generally accepted remedy has as yet emerged, however.

## Major aphthous ulcers (MjAU)

Major aphthous ulceration varies from the minor form in a number of important clinical details. In the first instance, the ulcers are generally larger than those of MiAU (Fig. 4.2) and they are much longer-lasting, up to a period of months in some cases. Probably as a result of the long periods of time involved, there is a marked tendency to the production of a heaped-up margin which, when a single ulcer is seen in isolation, may lead to the suspicion that the lesion is a malignant ulcer (Fig. 4.3). On eventual healing the ulcers may leave a substantial scar and this, together with the tissue destruction which may occur during the active phase of ulceration, may lead to gross distortion of the tissues (Fig. 4.4). MjAU produces lesions throughout the whole of the oral cavity, including the soft palate and tonsillar areas, and ulceration often extends to the oropharynx. The involvement of the posterior oral tissues is so characteristic of MjAU as to be diagnostic, even though the ulcers may initially be small. The depth of penetration of MjAU first led to the suggestion that the essential pathology was an abnormality of the underlying mucous glands; hence, the misnomer 'periadenitis mucosae necrotica recurrens'. It is now quite clear, however, that the basic pathological process involved in the production of MiAU and MjAU are the same and that involvement of mucous glands is a secondary process.

The numbers of ulcers present at the same time varies widely in MjAU from one to ten. Frequently, a single ulcer will persist for a long period, while other

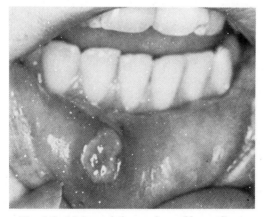

**Fig. 4.2.** Major aphthous ulcer of lower lip.

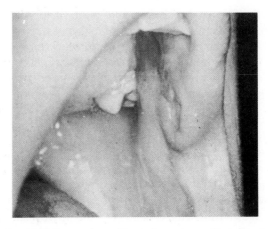

**Fig. 4.3.** Major aphthous ulcer of buccal mucosa with indurated margin.

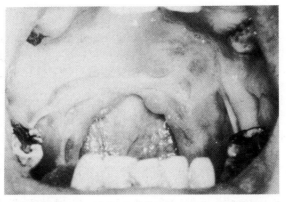

**Fig. 4.4.** Loss of soft palatal tissue following major aphthous ulceration.

(usually smaller) ulcers fade. As has been indicated, the size of the ulcers is variable, some being quite small (of the order of a few millimetres in diameter), whereas others (particularly of the persistent type) may be large (as much as 3 cm in diameter). Affected patients show a slight male preponderance and there is a wide variation in the age of incidence, although most patients experience their first ulcers in the first two decades of life. Unlike MiAU, there does not appear to be a cyclic pattern in MjAU and the ulcers are usually quite unpredictable in their onset. Long periods of remission may be followed by intervals of intense ulcer activity, without any evidence of the involvement of a precipitating factor. The prolonged and painful ulceration may present great problems to the patient, eating may become extremely difficult and the general health of the patient may suffer severely.

The histological findings in MjAU are essentially similar to those in MiAU, intense concentrations of mononuclear cells being found in relation to the area of ulceration. Immunofluorescent studies have also demonstrated close similarities between MjAU and MiAU; in both cases IgG and IgM are deposited in the cell membranes and intercellular spaces of the oral epithelium. Both circulating humoral antibodies and sensitized T-lymphocytes active against oral epithelium are found in MjAU, a feature which is much less marked in MiAU.

In view of the considerable distress caused to patients suffering from MjAU, it is particularly unfortunate that there is little to be done to give relief other than the adoption of the local therapy used in the treatment of MiAU. An exception to this generalization, however, lies in the use of systemic steroids, a form of treatment which must be approached with great caution. In order to abort a particularly severe attack of ulceration a short course of prednisolone, 40 mg daily, may be given for four days, reducing by 10 mg daily over the next three days. This may be effective in aborting the attack without the inherent risk of causing prolonged adrenal suppression. The remission gained by such therapy may last for some weeks or even months, but it is unlikely that anything resembling a long-term cure will result from such brief treatment. The hazards of long-term steroid therapy are such that, if the condition under treatment is not directly a danger to life, the treatment may, in the long-term, prove to be the cause of more problems than the original condition. The cases of oral ulceration which are actually life-endangering are extremely few and, hence, long-term systemic steroids are rarely considered as a reasonable form of treatment. A compromise form of treatment is the use of intralesional or sublesional injections of steroids, using either a needle or an instrument of the 'Dermojet' type. Triamcinolone acetonide suspension is most convenient for this form of therapy. However, there have been no fully documented reports on the efficiency of this form of treatment and, in the authors' hands, the results have proved variable. It is quite evident that such a form of therapy could be used only to control a single, long lasting lesion. Such lesions do occur, particularly in the oropharynx, and very occasionally such therapy may be of use in these circumstances, without in any way influencing the long-term behaviour of the recurrent condition. A further alternative to systemic steroids is the use of high concentration steroid mouthwashes. In some instances these can be used to replace higher dosages used systemically and thus reduce side effects. However, as pointed out in Chapter 1, absorption of the active agent takes place to a variable degree, and the use of this therapy must be limited and controlled in the same way as systemic steroid therapy. In general, patients on this form of therapy should be strictly supervised in a hospital clinic.

Other methods may be adopted to reduce the dosage of steroids necessary to control ulceration, azathioprine, for instance, may be used as 'steroid sparing' drug. However, the use of such aggressive therapy must be very limited. Some few other drugs are under investigation, mostly having an effect on the immune system.

The justification for steroid therapy must be considered carefully in certain medical conditions. For instance, those disorders which are normally considered a contra-indication to the use of systemic steroids include peptic ulceration, diabetes mellitus, thrombosis, hypertension, tuberculosis, and psychotic states.

## Herpetiform ulcers (HU)

This distinctive form of oral ulceration differs in many ways from both MiAU and MjAU, and it is a relatively less common condition. The term 'herpetiform' is not unlike that of dermatitis herpetiformis in dermatological practice, in that it refers to a morphological resemblance to lesions caused by *herpes* virus infections. However, it is an unfortunate one, causing constant confusion with *herpes* virus infections proper. In HU the ulcers are small (1–2 mm) and multiple (as many as a hundred ulcers may be present at the same time). Although any site on the oral mucosa may be involved, characteristically the affected sites are the lateral margins and tip of the tongue and the floor of the mouth (Fig. 4.5). The ulcers are grey and without a delineating erythematous border, making them be quite difficult to visualize. In spite of their small size, these ulcers are very painful and may make eating and speaking difficult. A single crop of ulcers may last for approximately 7–14 days, and the period of remission between attacks is widely variable; in some instances there may be repeated ulceration without any

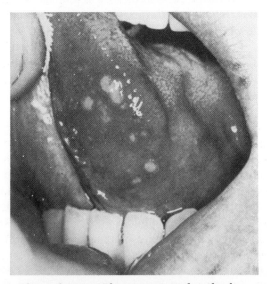

**Fig. 4.5.** Herpetiform ulcers of tongue. This patient was found to have coeliac disease and the long-lasting ulceration was eliminated by a gluten-free diet.

interval of remission. Where many ulcers are present there may be coalescence of several lesions to form a larger ulcer. The patients affected are predominantly female (2.6:1), the most common age of onset being in the 20–29-year-old group. Fortunately, this form of recurrent ulceration is relatively short-lived, most patients experiencing spontaneous remission within five years of onset. HU is the only form of ROU in which an early vesicular phase may occasionally be seen; such transient lesions rapidly break down to form the characteristic ulcers. When vesibular lesions are present they are most often seen on the gingival mucosa, presumably the physical characteristics of the affected tissues in this site are such as to allow of a somewhat more protracted preulcerative phase. Healing with scar formation has been described by Lehner, but other authors suggest that this is not a prominent feature of HU. It is certainly true that nothing resembling the gross tissue distortion caused by the scars of MjAU is seen as an aftermath of HU.

Whereas MiAU and MjAU represent differing intensities of the same condition with common histological and immunological features, HU shows quite different characteristics, there being no evidence of autoantibody formation or of T-lymphocyte involvement in the ulcerative process. The term 'herpetiform' as applied to this form of ulceration implies no more than a clinical similarity between HU and some viral infections, although there is little resemblance to an acute herpetic stomatitis. Histological and ultrastructural studies of HU have demonstrated the presence of intraepithelial vesicle formation before ulceration—a characteristic of viral infections—and also the presence of intranuclear inclusion bodies. There is, however, no evidence that these inclusions bear any relationship to the presence of viruses, and similar bodies have been demonstrated in the nuclei of epithelial cells affected by a wide range of conditions including keratoacanthoma, erythema multiforme, psoriasis, and lichen planus. Consistent viral infection of the ulcers has never been shown in culture experiments and virions have not been seen ultrastructural study of the lesions. Antiviral antibodies have not been demonstrated in the affected patients and, all in all, it is evident that the hypothesis of a viral aetiology for HU rests on somewhat slender foundations. However, no other more satisfactory explanation of the aetiology of HU has, as yet, been put forward.

The treatment of HU depends largely on one therapeutic aid, the tetracycline mouthwash. Steroids are quite ineffective in this condition. The response to tetracycline is often remarkably rapid and complete, the total effect being much greater than would be expected if the only action of the antibiotic was the elimination of secondary infection. There is, however, no evidence of any more fundamental action on the ulcers. In view of the pain and discomfort experienced by these patients, there is considerable pressure for the prescription of repeated courses of tetracycline, and it is often difficult to balance the short-term benefit of the treatment against its possible long-term disadvantages. Since, however, there is no other available method of treatment, it may be considered justifiable in some cases to accept the risks involved.

The relationship between MiAU and gluten enteropathy has been mentioned. It would seem that a similar relationship with HU exists; in fact, in the author's experience, many patients of this kind have ulcers of the herpetiform type. Again the necessity for screening procedures is evident.

## Mixed ulceration

Although the differences between the three basic types of ROU have been presented as being clear cut, this is not, in fact, the case. In many patients a long history of minor aphthous ulceration is followed by a transition period in which the ulcers become larger and longer lasting, until eventually the proper diagnosis is of major aphthae. Clearly, in the intervening period, the classification may be less than clear and at any given time in the same patient ulcers of both types may appear. Another form of mixed ulceration is evident in the considerable number of patients who suffer from both minor aphthous ulcers and herpetiform ulcers, often at the same time. This close association makes all the more remarkable the differences in behaviour of the two groups and, in particular, the responses to quite differing forms of therapy.

## ROU in children

Recurrent oral ulceration is not, by any means, a condition restricted to adults. The author has recently reviewed over a hundred patients with well established patterns of ROU from the age of seven years onwards. The initial appearance of the ulceration was, in the case of a number of patients, below the age of three years although these did not include cases of major aphthous ulceration. The first appearance of major ulceration in this group was at the age of seven years. The distribution of types of ulceration and the incidence of haematological abnormalities was entirely equivalent to that described in mixed age and adult groups. The onset of ulceration, contrary to frequently expressed ideas, did not seem to be, in any way, associated with puberty. Within this group there was only one patient with an abnormal jejeunal mucosa consistent with coeliac disease. This is consistent with the general finding that children with coeliac disease show evidence of this in their general condition and contrasts with the significant number of adult patients found to have a flat jejeunal mucosa associated only with oral ulceration and not with any more generalised illness.

## Beçhet's syndrome

Although ROU has been discussed as a group of conditions affecting the oral mucosa only, it is evident from a consideration of the associated immunological abnormalities that, in fact, complex bodily changes are involved and that the oral symptoms are no more than the end result of these. The presence of autoantibodies and of sensitized T-lymphocytes active against the oral epithe-lium has been well established in both MiAU and MjAU. The significance of the humoral autoantibodies is not entirely clear, since it seems that the precipitating factor in the ulceration is the increase in the concentration of immunologically active lymphocytes in response to some, as yet, unrecognized triggering mechanism. It is usually assumed that the presence of autoantibodies suggests a possible aetiological disease factor, but it has been suggested by Lehner that these may, in fact, represent a response to tissue damage during the ulcerative phase. Nevertheless, such autoantibodies do not develop during other ulcerative conditions of the oral mucosa and this would seem to indicate that they play a specific part in the disease process of ROU, although not of necessity a causative part. When autoantibodies are known to be produced as a result of tissue trauma these are transient, unlike those in ROU which are consistently present. This would seem to be a further source of evidence against the tissue damage theory of origin of these antibodies.

It has been shown that there is immunological cross-reactivity between the mucosal epithelia of the mouth, pharynx oesophagus, vagina, and conjunctiva as well as with the skin. This being so, it would seem reasonable to expect that the pathological process affecting the oral mucosa in ROU might also affect other immunologically similar tissues in the same way. Such is, in fact, the case and a number of conditions are recognized in which several such tissues are involved. The most usual complex of this nature is that in which oral ulceration is combined with genital ulceration and eye lesions. This is the condition which is usually referred to as Beçhet's syndrome although, in fact, in the fully developed condition there are more widespread lesions which may involve the nervous system, vascular system, or joints. The adoption of the term Beçhet's syndrome to describe the involvement of the mucosae alone is probably valid since it would seem almost certain that the syndrome represents a spectrum of activity ranging from, at the one extreme, uncomplicated oral ulceration to, at the other, a wide ranging disease process which involves vital structures and is potentially fatal. This wide range of activity may well reflect an equally wide range of specificity in the autoantibodies involved, the most specific being active against oral mucosa alone, the somewhat less specific being active against the antigenically cross-reactive genital and ocular mucosae. In the disseminated form of the disease the antibodies are, presumably, much less tissue specific.

In the more common restricted form of Beçhet's syndrome the oral ulceration may appear as any one of the three forms of ROU; indeed, the ulcers are quite indistinguishable from those of the uncomplicated condition. Genital ulcers are similar to those of the oral mucosa and may, if healing takes place with scar formation, lead to residual tissue loss and deformity similar to that in the oral cavity. The ocular involvement initially takes the form of anterior uveitis, a superficial inflammatory lesion of the anterior part of the eye, which may become more severe in later episodes and, perhaps, progess to involve other structures of the eye. This may lead to permanent damage by scar formation or, even, to blindness. The order of involvement of the oral and genital mucosa is variable, and the two may not be involved in ulceration at the same time; indeed, there may be a considerable interval between the involvements. Ocular involvement, however, is usually late, occurring sometimes after many years of intermittent oral and genital ulceration. In the generalized form of the disease, skin lesions of various kinds may appear, the most characteristic being papules which proceed to pustule formation. It is interesting to note that in some patients with skin lesions there is a marked skin reaction to trauma; if the skin is pricked a characteristic papular-pustular lesion results. This behaviour is reminiscent of the localization of oral ulceration by minor trauma in susceptible individuals. The neurological disease which may occur in these patients is the result of the appearance of centres of inflammation and necrosis within the central nervous system. The symptoms are variable, depending upon the location of the lesions, but in the early stages they may resemble those of multiple sclerosis. The vascular lesion is a vasculitis, perhaps complicated by thrombotic episodes, which may be either localized and minor or involve major vessels. The effect on the joints is that of a non-specific arthropathy. It has been suggested that major and minor criteria should be used to arrive at a diagnosis of Beçhet's syndrome. The major criteria are oral ulceration, genital ulceration, eye lesions, and skin lesions. The minor criteria include lesions of the nervous system, vascular system, joints, gastrointestinal tract, and pulmonary system. There is no agreement as to the number of types of lesion necessary to arrive at the diagnosis, but three major or two major together with two minor criteria are generally thought to be needed. It is the authors' viewpoint that, as this is a progressive condition, often beginning as simple oral ulceration without any other system involvement, it is impossible at any given time to differentiate with any degree of accuracy between simple oral ulceration and oral ulceration which might eventually proceed to oro-genital ulceration or more complicated states.

The figures quoted for the incidence of Beçhet's disease are difficult to interpret. The condition was first recognized in Turkey and was at first thought to be a disease of Mediterranean origin. However, a large number of cases have since been reported in Japan and, indeed, from other countries. It therefore seems possible that the current available figures refer more to the extent of investigation than to the true incidence of the disease. Similarly, various reports give widely differing values for the sex ratio of the patients involved. In some communications, the population is predominantly male, in others female, and in a third group of reports the figures are approximately equal. Again, this suggests an effect on the statistics of unrepresentative samples.

The local management of the oral lesions in Beçhet's syndrome is exactly as in the other forms of ROU and is similarly limited in effect. When generalized manifestations appear systemic steroids must be used, often for long periods of time, and during such treatment the oral ulceration is suppressed. Azathioprine may be used to reduce the dose of steroid necessary to suppress the lesions. On discontinuing such a course of therapy the oral ulcers tend to recur almost immediately, in contrast to the ocular lesions which may remain in abeyance for a long period of time. Thalidomide, cyclosporin A, penicillamine, cyclophosphamide, colchicine, and chlorambucil are among the drugs which have recently been reported as used to treat various aspects of the syndrome. The results have been variable and all have side-effects of considerable significance. Laevamisole, a drug mentioned with optimism in the last edition of this book has now been effectively abandoned in this context.

There have been many attempts to establish an immunogenetic basis for Beçhet's syndrome and for the forms of ROU discussed in this chapter. None have produced convincing results. In particular, the work on HLA associations in this group of conditions is confusing and contradictory.

# 5

# Lips, tongue, and saliva:
# facial swellings

SOME lesions which affect the lips and tongue merit separate description. Although many conditions which involve the oral mucosa in general may also affect the lips and tongue there are others in which these structures are specifically involved. Many of the lesions which involve the tongue depend on variations in the complicated surface structures, in particular the filliform papillae. The lips have no particularly remarkable structure, but they represent a transition area between the oral mucous membrane and the skin of the face, and may, therefore, take part in pathological processes affecting either. Because of their mobility both the lips and the tongue are particularly susceptible to trauma, either chronic or acute, and because of their cosmetic and sensory prominence even the most minor degrees of traumatic abnormality are quickly brought to the notice of the patient.

Disturbances of salivary flow and of taste sensation are included in the same chapter since these are often associated (or confused) with abnormalities of the tongue or of the lips. It is not unusual for a patient to complain of sensations of soreness or dryness affecting the mouth as a whole, but in particular the tongue and the lips. These patients may be difficult to diagnose and, in particular, it may be difficult to differentiate between symptoms which are essentially physical in nature and those which are psychologically induced.

## The lips

The most commonly troublesome lesion affecting the lips is probably angular cheilitis, a condition which has been discussed in Chapter 3 in relation to candidal infections. In most cases of angular cheilitis deep folds at the angles of the mouth become traumatised as a result of continual wetting by saliva. Secondary infection by *Candida albicans*, *Staphylococcus aureus*, or both, follows (Fig. 5.1). The angular folds are the result either of unusual anatomical configurations or, in the majority of cases, of unsatisfactory full dentures. It follows that the treatment should consist primarily of eliminating the abnormal tissue conditions at the site, antibacterial or antifungal therapy being used as a secondary measure.

In some few instances angular cheilitis is the result not of localized tissue

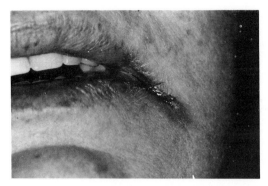

Fig. 5.1. Angular cheilitis with *Candida* present.

abnormality alone, but of generalized ill health. Angular cheilitis is regularly seen in patients with unsuspected iron, folate or B12 deficiency or with diabetes melitus. In other patients angular cheilitis may accompany xerostomia (dry mouth) as, for instance, in Sjögrens syndrome (discussed later in this chapter). Even more significantly, it may be an early marker of the onset of AIDS. There is no clinical difference between the lesions seen in these varying conditions and those described above as being the result of simple mechanical factors although, in the case of patients with generalized disease processes, the angular cheilitis is often an extention of a more widespread stomatitis. A systematic diagnostic approach must be used in these patients. Evidently, the definitive treatment of such patients depends on the results of investigations, but local treatment with antimicrobial agents may be symptomatically very helpful. It should always be remembered that angular cheilitis may be a multifactorial condition—a patient may have poorly fitting dentures as well as some underlying systemic defect.

A few other conditions may result in a more specific form of angular cheilitis. For instance, facial eczema may present at the angles of the mouth and mimic a simple angular cheilitis (Fig. 5.2). In this situation the diagnosis largely depends on the history of the patient; it is unlikely that a patient would initially present with eczema affecting the angles of the mouth only. The rapid response to local steroid applications (such as 1 per cent hydrocortisone cream) helps to confirm the diagnosis. In Crohn's disease angular cheilitis is a significant diagnostic indicator when associated with other markers of the disease (Chapter 8). This is a parallel condition to the anal fissures seen in this condition and, similarly, responds readily to local steroid therapy.

Less-frequently-seen lesions than angular cheilitis are vertical cracks in the lips; the most common site is in the midline of the lower lip. Such fissures are often remarkably resistant to conservative treatment and, in fact, surgical excision has been occasionally advocated for these more resistant lesions. The majority of these fissures are infected by *Staphylococcus aureus* and the most

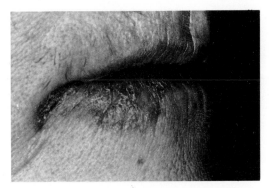

Fig. 5.2. Eczema resembling angular cheilitis.

satisfactory form of treatment consists of the elimination of the secondary infection (by the use of suitable antibiotic creams) followed by the use of steroid creams (1 per cent hydrocortisone with an antibiotic being most satisfactory). With such a regime most of these cracks can be persuaded to heal, at least temporarily. Should the predominant organisms involved not be a *Staphylococcus*, but *Candida*, then evidently the treatment should be modified suitably. However, the basic principle of the elimination of secondary infection followed by the use of a steroid application is the one most likely to be successful. Lip cracks of this kind show a marked tendency to re-occur and a permanent cure may be a matter of some doubt.

## Contact and actinic cheilitis

An irritation and scaling of the lips may result from sensitivity following contact with a wide variety of substances, and in particular the constituents of lipsticks. However, other medications used on the face, such as lip ointments, toothpastes, and, occasionally, foods may cause an allergic cheilitis. In general, the irritation is confined to the vermilion areas of the lip, but may extend beyond it in an eczema-like irritation of the perioral skin. This form of contact sensitivity is often difficult to diagnose and may require very careful history taking. For instance, a lipstick allergy may be caused by the minimal amounts transferred in kissing and so may remain quite unsuspected. In some instances the irritation may follow the action of sunlight on the lips. This may not be a primary actinic reaction, but may be secondary to a photosensizing effect of sunstances, such as those in lipsticks, which might in other subjects cause direct contact cheilitis. This is rather different from the severe hyperkeratotic reaction which may occur in some patients who work in full sunlight for many years; in the present example the exposure to sunlight may be minimal. In the case of a simple contact cheilitis treatment consists of tracing the sensitizing substance and eliminating it. Temporary relief may be given by use of topical steroids, 1 per cent hydrocortisone cream being satisfactory for this purpose. Actinic cheilitis may be prevented by the use of sun deflectant creams.

In a number of predominantly male patients long exposure to sunlight, often the result of outdoor work in a sunny climate, is followed by the onset of a cheilitis in which epithelial atypia may occur and which may progress to carcinoma. This is almost always most marked in the lower lip—the so called 'lip at risk'. Crusting and induration of the vermilion margin occurs, the induration being, in part, due to a fibrotic reaction of the connective tissues ('solar elastosis'; Fig. 5.3). Biopsy is necessary for a full assessment: the clinical appearance alone is an insufficient guide to the presence of worrying changes in the epithelium. If, on biopsy, marked atypia are found then excision of the vermilion area by a lip shave operation should be considered.

## Exfoliative cheilitis

Exfoliative cheilitis is a strange condition affecting only the vermilion borders of the lips and marked by the production of excessive amounts of keratin. This forms brown scales which may be spontaneously shed or may be removed by the patient (Fig. 5.4). It has been suggested that this condition is exclusively seen in female patients but several cases have recently been reported in males. The histology of the lesion is of a simple hyperparakeratosis and, so far as is known, this is in no way a premalignant lesion. There is no clear association with other abnormalities but some patients report an association with stress, the degree of keratin formation apparently increasing at times of increased tension. All systemic investigations in these patients prove negative and all forms of treatment have as yet proved unhelpful. Local and systemic steroids, cautery, cryosurgery, and many other forms of treatment have been attempted without success. In those with an apparent stress-related condition, intermittent mild tranquilization has been reported as helpful although the mechanism for this is

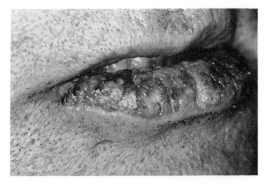

Fig. 5.3. Solar keratosis of the lower lip.

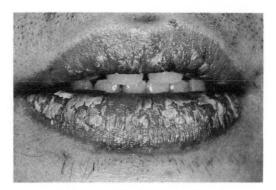

**Fig. 5.4.** Exfoliative cheilitis.

far from clear. Eventually, this condition spontaneously resolves, but whilst present it is sufficient to cause the patient considerable distress.

## 'Chronic granulomatous cheilitis' (orofacial granulomatosis)

There exist a group of conditions in which swelling of the lips is associated with the production of chronic granulomatous lesions of the labial and buccal mucosa. Clinically, these lesions bear a remarkable resemblance to those seen in patients with Crohn's disease, but all investigations may prove entirely negative so far as other parts of the gastro-intestinal tract are concerned. The patients are predominantly young and present with marked swelling of one or both lips. Associated with this swelling there is marked painless enlargement of the submandibular lymph nodes and an angular cheilitis. The buccal and labial mucosa may be oedematous and corrugated and may occasionally become ulcerated. If the affected areas are biopsied, the histology shows non-caseating, tuberculoid, or sarcoid-like lesions within the corium. However, in these patients the Kveim test is usually negative and tubercle organisms cannot be demonstrated. Quite often the extent of the oedema of the lips is quite disfiguring although there is very little irritation except from the angular cheilitis and from vertical cracks which may occur (Fig. 5.5).

The true nature of this condition is not known, but it has recently become plain that Crohn's disease is not in fact, as was once thought, confined to the ileum and it is probable that the lesions of 'granulomatous cheilitis' represent a localized form of the same pathology. The Melkersson–Rosenthal syndrome is described as having similar features with an associated seventh nerve paralysis and a deeply fissured tongue as other components of the syndrome. The contemporary view is that lesions of this type, together with those of oral Crohn's disease and sarcoid probably represent a spectrum of disease processes, and that there is a merging of features which may make a clear cut diagnosis

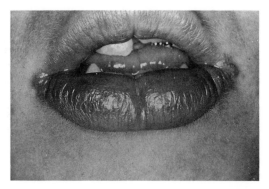

**Fig. 5.5.** Swollen lip in a patient with both oral and lower gastrointestinal Crohn's disease.

difficult in many cases. It is the author's strong view that patients with these features should be considered, in the absence of any more specific diagnosis, as suffering from a localized form of Crohn's disease and, hence, at risk of developing lesions in other areas of the gastro-intestinal tract. These patients and their management will be further discussed in Chapter 8.

## Angio-oedema

This term is used to describe two rather different conditions. In one (also known as angioneurotic oedema) it refers to an acute non-specific allergic reaction to a variety of stimulae, such as foodstuffs or antibiotics, which results in widespread and often dangerous soft tissue swelling. It often involves the facial, oral, and pharyngeal tissues. This is a form of acute urticaria depending on the presence of specific antibodies to IgE which bind to receptor sites containing IgE in the walls of mast cells. This leads to degranulation of the mast cells, and consequent liberation of histamine and histamine-like substances. Vascular permeability, with consequent oedema is the result. This condition may be a very urgent one—respiratory occlusion is a real hazard—and must be treated as such using the usual measures to combat acute atopic reactions such as intravenous steroids, antihistamines, and adrenaline.

The second condition, more often described as hereditary angio-oedema, is a more specific condition which is genetically determined (autosomal dominant) and in which C1 esterase inhibitor is deficient. C4 levels are also low. As a result of the C1 esterase inhibitor deficiency the action of the complement system is uncontrolled and activation of kinins leads to capilliary permeability and consequent oedema. In its fully developed form this is a serious, and indeed, life-threatening condition involving the facial tissues and the airway. Minimal local trauma, as in dental treatment, may precipitate an attack. Treatment of the acute episodes has been carried out with fresh plasma or plasminogen activation inhibitors such as tranexamic acid. A currently recommended long-term treatment is with danazol, a drug primarily active in controlling pituitary activity, but which is also effective in raising C1 esterase plasma levels. Fortunately, the major condition is rare, but there exist a number of less well marked variations in which the swelling is less marked and the immunological changes less characteristic, although measurable. These are often very difficult to recognize and to treat.

In both these conditions the initial swelling very often affects the lips and may be mistaken for a local dental problem.

## Perioral dermatitis

This relatively uncommon condition is often a considerable diagnostic problem and the patients are from time to time seen in the Oral Medicine clinic for investigation. The patients are almost always young adult females and complain of an erythematous rash on the facial skin around the mouth (Fig. 5.6). In some cases, but not all, there are papules present. In some patients there is a history of the use of steroid creams on the face, but this is not by any means invariable.

The diagnosis is essentially clinical—there are no systemic changes detectable. Many of those patients with papules present respond to a long-term, low dose regime of tetracyclines, much as in acne. Paradoxically, some patients respond well to low potency steroid creams, such as 1 per cent hydrocortisone. As in all cases where the facial skin is involved caution must be exercised in the use of more potent steroid preparations because of the possibility of atrophic changes. In view of the wide range of presentation and reaction to therapy it is possible that perioral dermatitis represents a reaction pattern rather than a single entity, and that more than one basic aetiological process may be involved.

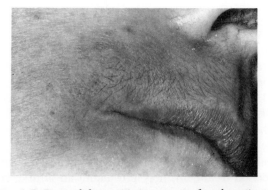

Fig. 5.6. Perioral dermatitis in a young female patient.

## 'Lick eczema'

This is a further perioral condition which often causes diagnostic confusion. The patients are young—often children—and complain of a sharply delineated zone of irritable, scaly skin around the mouth (Fig. 5.7). This condition, as its name implies, is simply the result of a licking habit which the patient may strongly deny, even when actually performing the licking action during the examination! The treatment is to stop the habit. If this is done the lesion rapidly fades. However, such a well ingrained habit (of which the patient may be entirely unconcious) may be difficult to eradicate. A removable appliance with a rough or sharp edge to interfere with the tongue action is often a very successful aid to this.

## Cheilo-candidiasis

There are a number of cases reported in which the lips become the site for a heavy candidal infection. These lesions are reported as occuring bilaterally, predominantly on the lower lip, and appear as ulcerated granular areas from which *C. albicans* can be freely cultured. This is reported to occur in generally healthy patients, but there are strong suggestions of prior local abnormality which might lead to a secondary candidal infection, for instance, solar irritation in some patients reported from Australia. In some cases the cheilitis has been associated with intra oral chronic candidiasis. These lesions have been considered by several authors to represent candidal infection affecting intrinsically unstable epithelium and it is suggested that, as in chronic hyperplastic candidiasis (candidal leukoplakia), early treatment by antifungals might lead to resolution, whereas delay might lead to increasing epithelial dysplasia.

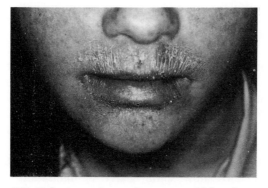

Fig. 5.7. Lick eczema in an eleven year old female patient.

## The tongue

The tongue may be involved in a wide range of disease processes which affect the oral mucosa in general. Apart from this involvement in generalized oral conditions there are a number of lesions which are quite specific to the tongue. Many of these depend on changes in the specialized epithelial covering of the tongue and, in particular, in the filliform papillae; these structures appear to be particularly susceptible to changes brought about by systemic abnormalities. For instance, in anaemias (in which the oral mucosa as a whole may undergo some change) the predominant oral abnormality may be of the tongue. When a lesion of any kind is present on the tongue it may be brought rapidly to the notice of the patient because of the mobility of the organ and its rich supply of sensory nerve endings. This, however, is by no means invariably the case and there are many examples of lesions (for example, the widespread ulceration of major erosive lichen planus) in which the patient complains of far less pain than might be expected. Of greater significance is the fact that neoplasms of the tongue may grow to a considerable size before secondary infection leads to pain symptoms; an early carcinoma of the tongue may be quite painless.

The normal gross structure of the tongue includes fissures. There is a very great variation in the number, depth, and the arrangement of these, and most are entirely anatomical. There is considerable doubt as to the extent to which the pattern of fissures may change during disease processes although it does seem that in a few conditions such as (for example) chronic candidiasis of long standing, the fissures appear so exaggerated and of such abnormal form that they must be considered as part of the pathological change (Fig. 5.8). In general, however, there is little evidence of changes in fissure pattern due to less exaggerated pathologies. There is no doubt, however, that fissures of normal form and distribution may play a part in the modification or, even, the initiation of pathological processes since conditions of anaerobic stasis are present in the depths of these structures and present opportunities for unusual bacterial growth. This being so, it is common to find superficial infections and similar conditions concentrated at and alongside the fissures. Similarly, in a situation which may lead to a sore and irritated tongue, the fissures may be the first and the most severely affected. Often the patient attributes the fissuring to the condition which is causing the pain but usually the converse is true.

Because of its mobility the tongue is particularly susceptible to trauma which may be acute (as following the fracture of a carious tooth) or in a more chronic form as, for instance, by the adoption of a habit of rubbing the tongue over the edges of the anterior teeth, a denture, or an accumulation of calculus. The lesions produced vary greatly, from mildly erythematous patches often seen on the tip of the tongue, to frank ulceration. These traumatic ulcers are often extremely painful and, although they may be evidently related to the sharp edge in question, they may show some degree of induration due to localized inflammatory changes together with whiteness of the surrounding epithelium,

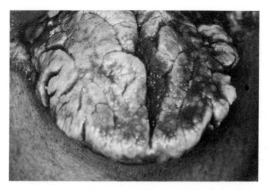

Fig. 5.8. Deep fissures of the tongue in an adult with chronic mucocutaneous candidiasis.

changes which arouse immediate suspicion of malignancy. However, it has been amply demonstrated that superficial ulcers of the tongue, surrounded by a white margin, are virtually always benign and of traumatic origin (Fig. 5.9). The easiest way to allay these suspicions is to remove the offending tooth or appliance, or to deal with it in some other suitable way in order to eliminate the trauma completely. Traumatic ulcers resolve very rapidly and virtually all trace may be expected to have disappeared within a week of removing the irritant. If this resolution has not occurred in the expected time then the lesion must be viewed with suspicion and biopsy carried out. In the tongue, as in other areas, there is little evidence of a relationship between physical trauma and malignancy, but the possibility must evidently be considered.

More difficult to diagnose are the more chronic forms of irritation resulting

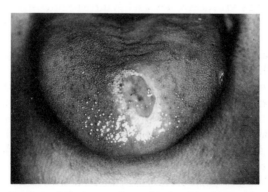

Fig. 5.9. A traumatic ulcer of the tongue brought about by friction from an orthodontic appliance.

from habits of various kinds, often resulting in no very obvious lesions, but often causing the patient a great deal of concern. In these cases it may be difficult to decide whether any systemic factor is involved in the production of the symptoms described by the patient and it may be necessary to carry out a full series of blood investigations before coming to the conclusion that a habit is the only factor involved. In particular, there are a number of patients in whom the irritation of the tip of the tongue and often also the lower labial mucosa follows on the most minor change in morphology of the anterior dentition, say by the replacement of a crown or the insertion of a replacement partial denture. Frequently, there is no discernable abnormality in the restoration to account for this; none the less it is evident by the mild erythema of both the tip of the tongue and the mucosa of the lower lip that trauma is occuring. It is often difficult to help such patients in the absence of any discernible fault in the restoration. An all over soft appliance covering the teeth affected by the restoration for a short time has been found the most helpful in the authors experience.

## Coated and hairy tongues

The tongues of all normal individuals have a coating consisting of a layer of mucous, desquamated epithelial cells, organisms, and debris. In the healthy individual the tongue is mobile, there is a rapid flow of saliva and this coating is kept to a minimum. However, with the slightest disturbance to the health of the individual the balance is upset and the coating may quickly become very much thicker. A lack of mobility of the tongue which may be caused by the most minor painful lesions, a disturbance in saliva flow, an excess of tobacco or of alcohol, a gastric or respiratory upset, a febrile condition, or one of a wide variety of other disturbances may result in a build-up of the tongue coating sufficient to produce a white or coloured plaque. The colour of such a coating depends on a variety of factors, such as tobacco usage and dietary habits, and is of very little diagnostic significance. However, such a coating, particularly in circumstances of unusual stasis (such as in the treatment of facial fractures by intermaxillary fixation) may be unpleasant to the patient and its removal may be quite difficult since very often it is firmly adherent to the tongue below. Effervescent mouthwashes contain oxidizing agents, for instance a dilute solution of hydrogen peroxide, may be effective, as also may be mouthwashes containing ascorbic acid. This substance is a mucous solvent and is quite effective in loosening and detaching coatings from the tongue.

　　In hairy tongue the lesion does not consist simply of a coating on the surface of the tongue but represents an elongation of the filliform papillae, often to many times their original length (Fig. 5.10). With this elongation the papillae often take on a dark colour, black or brown being common (Fig. 5.11). The mechanism of the formation of these coloured hairy tongues is quite unknown, there apparently being many initiating factors. For instance, hairy tongue frequently follows a course of antibiotic therapy and may resolve quite rapidly

**Fig. 5.10.** Elongated filliform papillae in 'hairy tongue' (scanning electron micrograph).

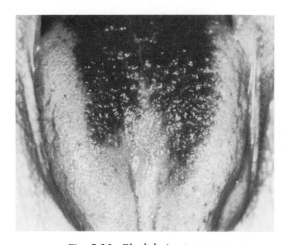

**Fig. 5.11.** Black hairy tongue.

on the completion of this course of treatment. Other hairy tongues apparently appear completely spontaneously and no cause is ever found for them. Equally doubtful is the source of the pigment involved. It is usual to relate this to pigment-producing organisms entrapped within the papillae, but in fact, no such organisms have ever been demonstrated. In the past, the presence of hairy tongue was often ascribed to candidal infection, but again, it has never been shown that there is any true association between candidiasis and the production of the elongated papillae.

Treatment of hairy tongue is remarkably difficult. Those cases associated with antibiotic therapy frequently resolve at the end of a course, but not invariably so. The use of effervescent and mucous-solvent mouthwashes may be helpful in reducing secondary irritation and thereby producing suitable conditions for the resolution of the abnormality, but again, the results are variable. Vigorous brushing of the tongue is often advocated, but this is often found to be most unpleasant by the patient. Chemical cauterization of the elongated papillae by substances such as trichloracetic acid has been advocated in the past, but this may be painful to the patient and has not been shown to be particularly effective. Cryosurgery has not proved superior to chemical cautery. Some claim success with the use of antiseptics such as chlorhexidine mouthwashes. It would seem reasonable to suggest that such treatment, if effective, must operate by the removal of secondary infection from the affected papillae. These mouthwashes and antiseptic preparations must, however, be approached with some reserve, since most of them have been from time to time implicated in the production of hairy tongue rather than in its cure. Altogether this is a most unsatisfactory condition to treat both in the lack of knowledge of its aetiology and because of the absence of a uniformly successful remedy.

## Atrophic lingual epithelium

At the opposite extreme from the proliferation of the filliform papillae seen in hairy tongue are the atrophic changes in the papillae found in a number of generalized diseases, but in particular, in haematological and nutritional abnormalities. Atrophic change in the oral epithelium as a whole may be the consequence of a wide range of abnormalities (for instance, iron deficiency, latent iron deficiency, megaloblastic anaemias, and nutritional deficiencies of various kinds). These atrophic changes are sometimes, but not invariably, found predominantly in the filliform papillae and the surface of the tongue may become red, shining, and painful. In the past, characteristic appearances have been associated with specific conditions, but it is now accepted that a wide spectrum of generalized abnormalities may be responsible for the production of similar epithelial changes. Conversely, a wide range of oral signs and symptoms may occur in different patients with similar conditions. It is certainly true, however, that all patients with painful, depapillated, or reddened tongues should undergo haematological investigation. These investigations should include estimations of serum iron, total iron-binding capacity, saturation, $B_{12}$ and folate, as well as the more usual haemoglobin estimation, and full blood count. Many abnormalities are shown only after such detailed examination and not by a simple blood-screening procedure.

A sore tongue may be found in association with many generalized diseases, particularly when the blood picture is secondarily affected or the saliva flow disturbed. For instance, in rheumatoid arthritis, where there may be a secondary anaemia, it is not unusual to find a sore tongue. Similarly, in Sjögren's syndrome, in which the salivary flow is reduced, the tongue may become dry, reddened, or painful. There are, however, some patients in whom no cause can be found for the atrophic changes. In older patients the atrophy may be regarded as an age change, but it is not clear to what extent this may represent subclinical systemic changes. In many conditions in which depapillation has taken place the original morphology of the tongue epithelium does not return with healing. For instance, in major erosive lichen planus a large area of the tongue may be ulcerated and the healing process may proceed with the formation of flattened epithelium virtually devoid of filliform papillae.

Apart from the patients in whom a recognizable pathological condition exists, there is a group, largely of older female patients, who complain of tenderness and soreness of the tongue, and who show no obvious signs to account for this. All tests may show negative results and it is tempting to classify such complaints as being psychogenic in origin. There is no doubt that in some patients this may be so, but the possibility of habit-induced irritation must always be considered in such cases.

## Superficial glossitis: midline glossitis

It has been pointed out in Chapter 3 that the tongue is often involved in candidal infection. This is particularly so in the chronic infections dependent on immuno-logical defects (such as chronic mucocutaneous candidiasis) or in debilitating chronic diseases, and in such cases the tongue may not only be covered by plaques of candidal pseudomembrane, but may also become deeply fissured and apparently fibrosed. It should, perhaps, be repeated that candidiasis of the tongue, as in the oral cavity as a whole, is the mark of some abnormality in the health of the patient. Glossitis caused by other identifiable organisms, either superficial or deep, is very rare and, again, practically all cases are secondary to some severely debilitating disease. In conditions of relative good health, how-ever, such infections are virtually unknown.

More common, and of great interest, are the lesions of chronic superficial glossitis usually seen in apparently healthy patients. These are virtually always in the midline of the tongue in sites varying from immediately in front of the vallate papillae (where the condition is known as median rhomboid glossitis because of the characteristic shape of the lesion in that site) forward over the dorsal surface of the tongue. Median rhomboid glossitis has for long been considered to be a developmental abnormality in some way connected with the site of the embryonic tuberculum impar, but it has recently been shown quite convincingly by Cooke that this lesion is similar to those occurring more anteriorly on the tongue. These consist of rounded depapillated patches, frequently standing above the general surface of the tongue and with a red, white, or yellow appearance. Cooke has suggested that these lesions of midline glossitis are associated with candidal infection of the epithelium and that treatment with antifungal antibiotics, if sufficiently protracted (several months may be necessary) will considerably reduce the area of the lesions. Not all lesions are fully resolved by this treatment and cryosurgery has been successfully used to eliminate residual abnormalities. It need hardly be said that any doubt as to the nature of the condition should be resolved by biopsy examination. In general, however, the characteristic appearance, site, and texture of the lesion is such as to enable a confident initial diagnosis to be made on clinical grounds (Fig. 5.12).

## Geographic tongue

Depapillation of the tongue is also a marked feature in the condition usually described as geographic tongue or erythema migrans. In this condition the pattern of depapillation is quite characteristic, the affected areas occurring as red patches surrounded by white borders. The patches are distributed over the surface of the tongue in a map-like fashion (hence the name geographic tongue) and tend to vary in their sites, apparently moving about the surface of the tongue, hence the name erythema migrans (Fig. 5.13). The appearance of the lesions may be the only complaint of the patient although very often there is a tenderness of the tongue to spicy and well seasoned foods. This condition often proves extremely worrying to the patient in view of its often quite spectacular appearance. The basic aetiology of this condition is quite unknown; it is rarely associated with any underlying systemic disorder although, should there be any coincidental condition such as anaemia, there may be an increase in the tenderness of the tongue. Diagnosis is in general simple, depending entirely on the appearance of lesions, although it may in some cases be necessary to see the patient on several occasions to confirm the migratory nature of the condition. In

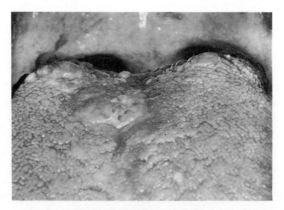

Fig. 5.12. Median glossitis.

Fig. 5.13. Geographic tongue (erythema migrans).

some patients the lesions may be somewhat less well defined and with a less marked white border although, in practically all cases, at least a small area of white demarcation may be seen. In other patients the lesions may occur in a more static manner or in a single site and this may cause confusion with a simple traumatic lesion (Fig. 5.14). However, there are a few cases of geographic tongue which cannot eventually be diagnosed on a clinical basis alone. Biopsy is rarely necessary to confirm diagnosis, but evidently should immediately be

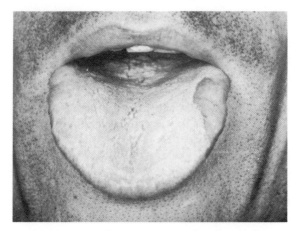

Fig. 5.14. A solitary area of erythema migrans.

considered if there is a possibility of a more aggressive lesion. No successful treatment is known for geographic tongue.

The age range of the patients involved is very wide. Although geographic tongue is generally considered to be a condition affecting adults the author has seen patients from the age of 3 years upwards. The association of geographic tongue with psoriasis is a controversial matter; there are no statistically based reports and, in the case of two commonly occuring conditions, the question of simple coincidence arises. A further suggested association is between geographic tongue and deep fissuring. Again, there is no clear evidence that this is other than a random association of two relatively common conditions.

In the context of psoriasis it is interesting that there has been a recent report of enlargement of the fungiform papillae in association with psoriasis of acute onset; this is an oral finding which had not previously been reported.

## Burning mouth

A number of patients present with the complaint of a generalized soreness or burning sensation in the mouth, often particularly affecting the tongue. Although, evidently, some such patients may have readily diagnosable conditions to account for this, others may show no recognizable abnormality of the oral mucosa. These are the patients with the so-called burning mouth syndrome (although this is by no means a syndrome as normally defined).

The assessment of these patients is a difficult procedure—denture induced problems, systemic disease, and dry mouth must all be eliminated. A proportion of the patients will, at some time, have had the problem ascribed to an allergy to denture materials and it may be extremely difficult to dissuade the patient from this attractive, but exceedingly unlikely view of the situation. In recent surveys the patients with this kind of complaint were predominantly (but not exclusively) female with a mean age of around 60 years and wearing complete dentures. The majority of patients complained of a burning sensation in the tongue and upper denture bearing area. The next most common site was the mucosa of the lips and, then, other sites on the oral mucosa.

From the results of these surveys it would appear that three groups of patients emerged: those with a demonstrable source of local irritation (50 per cent), those with an identifiable systemic abnormality (30 per cent), and those with a psychogenic background (20 per cent). The systemic abnormalities included haematological deficiencies and unsuspected diabetes, but contrary to previous unconfirmed views, female endocrine abnormalities were not prominent. The causes of local irritation were predominantly to do with faulty denture design, but also included excessive smoking and the frequent use of mouthwashes. On the whole, patients appear able to recognize denture problems as a cause of burning mouth, but it must be recognized that some patients are quite intolerant of dentures, however well constructed.

Quite clearly, both careful clinical examination and the use of appropriate screening procedures play a vital part in the diagnosis and management of these patients. The significance of abnormal haematological findings will be further discussed in Chapter 9.

## Disturbances of taste

Patients complaining of disturbance of taste sensation are among the most difficult to diagnose. There is often nothing other than the patient's own description to characterize the condition and objective tests may show very little. Physiological tests do determine the patient's ability to differentiate the basic taste sensations of sweet, sour, bitter, and salt may be carried out, but these in themselves may give little information, since the sense of taste as a whole depends on a mixture of associations of taste and scent. True neurological disturbance of the sense of taste is extremely rare, but may occur in some generalized abnormalities of the central nervous system, particularly degenerative diseases. It may also occur in late syphilis. A somewhat more common (although still rare) neurological disturbance is that due to surgical trauma to the chorda tympani following operations on the middle ear. If the nerve is disturbed either during operation or by post-operative oedema an area of taste disturbance may develop on the side of the tongue corresponding to the terminal distribution of the chorda tympani with the lingual nerve. In a few patients with Bell's palsy (Chapter 11) there is a similar loss of taste sensation due to involvement of the chorda tympani on the affected side.

The most common cause of a disturbance of taste sensation or of a foul taste in the mouth is the presence of a pyogenic infection, most commonly the result of periodontal disease, infection in the nose or sinuses, or a discharge from a lesion such as an infected dental cyst. Such an infective focus may be very difficult to detect, particularly when present in a relatively inaccessible area such as a salivary gland. It is evident that the clinical examination of such a patient should always include a careful search for pus-contaminated saliva from any of the major salivary glands as well as radiographic assessment for the presence of possible intrabony infected lesions, sinusitis, and similar possible infective foci. Clearly, such patients are suffering not from a true disturbance of taste sensation, but from the chronic imposition of unpleasantly tasting material into the oral cavity. This is often a very difficult differentiation for patients to make.

Disturbance or loss of taste may occur in a number of generalized diseases such as diabetes and the anaemias, in particular pernicious anaemia. In some instances of pernicious anaemia the disturbance of taste may precede the onset of characteristic haematological change, and the basic aetiology will be determined only by a more rigorous haematological examination than a simple blood count. The same scheme of investigation should be adopted as that suggested for the investigation of the sore or depapillated tongue. It is certainly true that many patients with disturbances of taste and sore tongues are now found to have some basic underlying haematological abnormality when previously they would have been regarded as being physically fit patients. In addition to the haematological investigations a urine examination should be carried out for the presence of sugar. However, when all such examinations have been performed there remains a group of patients in whom no abnormality of any kind is detected, but who continue to complain of an unpleasant taste. These patients are often obsessional, but it is very difficult to determine whether this is the cause of their complaint or the consequence of it. Some of the most difficult patients to treat among those with disturbances of taste sensation are those who are wearing relatively recently acquired dentures. Few taste buds are covered by these and the olfactory nerve endings are in no way affected, but none the less, some such patients complain of an abnormality or even complete loss of taste sensation. The explanations for this are not particularly convincing, but it would seem that variations in texture between the uncovered palate and the denture may play a part.

The dentist's responsibility in such cases as those described above is to eliminate by detailed examination any of the local or generalized causes mentioned. If this examination is negative and if all concerned are sure that there is no nasal or similar aetiology, then there is little point in attempting treatment. Mouthwashes may occasionally help, but in general the patient obsessionally concerned with a persistent taste is completely resistant to local treatment.

## Disturbances of salivary flow

### Xerostomia

Dryness of the mouth is a common complaint which may have many different causes. However, most patients with a demonstrable lack of saliva may be shown either to have loss of secretory tissue in the salivary glands or a disturbance in the secretory innervation mechanism brought about either by neurological disease or by drug action.

There are conflicting reports of the effect of age on salivary secretion, but it is generally accepted that histological changes occur within the salivary glands with increasing age. The most important change from the point of view of secretory activity is the replacement of functional units by fat and connective tissues. Clinical surveys have shown widely differing results, but there seems no doubt that in certain patients in whom no generalized disease is found, increased age is accompanied by a decrease in salivary flow. These patients (as many others with reduced salivary secretion depending on other factors) complain of dryness of the mouth and often also of generalized soreness of the mouth. In edentulous patients there may be instability and lack of retention of dentures. However, some workers believe that enough potentially secretory units are present in the salivary glands of healthy patients of all ages to produce adequate salivary secretion. Clinical trials are currently taking place to assess the effect of drugs such as pilocarpine in stimulating salivary flow in these circumstances.

Clearly, such stimulatory therapy would be of no use if the xerostomia were due to total replacement of secretory cells.

A high proportion of patients complaining of xerostomia are shown after investigations to have some systemic factor responsible for the reduction in salivary function. These systemic factors may be associated with a wide range of disease processes (for example, anaemias and endocrine disturbances), but many of these patients are suffering from the Sicca syndrome of Sjögren's syndrome. These conditions will be discussed in more detail below.

Although neurological disease, either central or peripheral, may be responsible for a decrease in the secremotor stimulation of the salivary glands and, hence, the dryness of the mouth, the most common cause for this is the action of drugs. A very wide range of drugs have been implicated in this way, the site of action being either central or in the autonomic pathways. Many groups of drugs have been described as having this kind of action including antihistamines, antihypertensives, and sedatives. However, the more common drugs which cause xerostomia at the present time are the psychotropic drugs and, in particular, the antidepressants and tranquillizers.

The use of saliva substitutes may be of some help, but as is pointed out in Chapter 2, their effect is relatively transient. There is no doubt that xerostomia is a difficult condition to treat.

## Excessive saliva

Excessive saliva (sialorrhoea) is most commonly complained of by patients wearing dentures for the first time. Any other foreign body in the oral cavity can have a similar effect and, in fact, one of the commonly used methods of stimulating salivary flow for experimental purposes is to use an inert foreign body within the mouth. Most patients eventually become used to the new dentures and during this process the excess salivary flow usually disappears. In a few patients, however, this may prove an intractable problem. Apart from foreign body stimulation, many infected or ulcerative lesions in the mouth may temporarily cause an increase in salivary flow which adds to the discomfort of the initial condition. A similar effect is often seen in carcinoma of the mouth in which the increased salivary flow (often associated with loss of function) results in constant dribbling of saliva, a particularly unpleasant side effect of the neoplasm. Very few modern drugs induce excessive salivation: the most common are anticholinesterases as used in the treatment of myasthenia gravis.

In a few patients complaining either of excessive or of reduced salivation there are no obvious symptoms and no abnormality is ever detected. This situation is presumably the result of mild psychiatric disturbance and indeed the patients are often obsessional in their nature. No more than symptomatic treatment can be given in such a situation and, in view of the outlook of the patients concerned, this may not be very satisfactory.

Treatment for excessive salivation depends largely on the elimination of (or habituation to) the causative factor, whether it be a foreign body or an infective lesion. The use of drugs to suppress salivary flow is rarely indicated since virtually all drugs with a marked salivary suppressive effect also exert other, and often more significant, effects.

## Salivary gland disease

A number of conditions affecting the salivary glands may more properly be considered as being surgical problems. These include such lesions as the cysts and neoplasms arising in the glands and also conditions arising as a result of the presence of calculi or other obstructions in the glands or ducts. However, with the elimination of these two groups of important salivary gland diseases there remain a number of conditions which should be considered as within the scope of oral medicine.

The parotid glands, lying partially concealed by the ascending ramus of the mandible, are not particularly easy to palpate. Tenderness and swelling are best detected by standing in front of the patient and by placing two or three fingers over the posterior border of the ascending ramus of the mandible. Backwards and inwards movement of the fingers with light pressure is almost always all that is needed to detect tenderness in the superficial part of the parotid. This manoeuvre is necessary to differentiate parotid tenderness from that of the temporomandibular joint or the masseter with which it is often confused. It must also be remembered that the painful signs resulting from temporomandibular joint; muscle dysfunction may extend to the upper pole of the parotid with which the joint is in close anatomical relationship. When examining a parotid gland the duct papilla must also be examined intraorally for signs of inflammatory change. The saliva can best be assessed for apparent abnormality by lightly compressing the skin overlying the duct with the fingers. If the cheek is held retracted the saliva expressed by this manoeuvre will be seen coursing downwards over the buccal mucosa from the duct papilla. Subjective assessments of the flow rate of saliva made in this way are, however, completely unreliable.

The submandibular gland may be felt below the angle and body of the mandible, this simple palpation being reinforced by bimanual palpation with a finger in the floor of the mouth, gentle pressure being exerted between the examining hand (below the mandible) and the finger. As in the case of the parotid gland, the submandibular duct should be observed for signs of inflammation and a subjective assessment made of the quality of the saliva. Just as is the case of the parotid gland it is difficult to specify the 'normal' palpation features of the submandibular gland. In some entirely normal individuals it is possible to palpate the gland, while in others it is not.

## Sialadenitis

Sialadenitis is the term used to describe inflammation of salivary glands, most commonly the result of viral or bacterial infection, but occasionally due to other causes. Bacterial sialadenitis is usually a secondary consequence of obstruction of the duct of the gland and is relatively rarely the primary pathological process whereas viral infections frequently affect previously normal salivary glands. Generally speaking, sialadenitis is a clinical problem affecting the major salivary glands. However, it may also occur in minor salivary glands, either as a primary phenomenon (as in Sjögren's syndrome) or as a secondary feature of some other condition such as pipe smokers palate in which the palatal salivary glands often become inflamed and enlarged (see Chapter 7).

Of the viral infections of the salivary glands by far the most common is mumps which most commonly affects the parotid glands, but also from time to time involves the other major glands. The mumps virus is transmitted by direct contact or by droplet infection, and a clinical infection is heralded by a feeling of malaise and often by abdominal pain. Following this, there is swelling of one or both parotid glands, sometimes associated with a swelling of the submandibular glands also. In many cases, one of the parotid glands is by far the more severely affected. With the swelling of the gland there is pain and tenderness together with a general malaise and a fever. The swelling of the salivary gland is at its maximum within 1 or 2 days of the onset of symptoms and is followed by a gradual resolution over 1 or 2 weeks. Mumps often occurs in minor epidemic form and is usually recognized on clinical symptoms alone. Confirmation may be obtained by antibody measurements although these are rarely used in practice. Electron microscopy (if available) will immediately identify the characteristic virus. Treatment of mumps is entirely symptomatic, with the use of antipyretics and mild analgesics. Isolation is evidently important since the disease is very contagious.

A well established complication of mumps is the onset of orchitis or oophoritis. These conditions are caused, not be direct viral infection, but by the production of autoantibodies directed against the tissues of the reproductive organs. After an attack of mumps immunity is usually complete and, therefore, a recurrent swelling of the parotid region is almost certainly not due to a recurrence of mumps. It should be remembered, however, that salivary gland infections may be caused by other viruses although these are relatively rare in comparison to mumps.

As has been pointed out, bacterial infection of the salivary glands is commonly secondary to an obstructive condition such as calculus formation or duct stricture. This being so, it is evident that a single gland is normally affected; most often this is a parotid gland. In this condition the gland is painful, tender, and swollen, and pain is radiated to the ear and the temporal area. Intra-orally, the duct of the affected gland may be seen to be swollen and reddened, and the duct papilla enlarged. Purulent saliva may be milked from the duct by manual pressure through the substance of the cheek.

Treatment of an acute parotitis tends to be somewhat protracted, presumably because of the low rate of excretion of antibiotics in the saliva. It is evident that bacterial swabs should be taken and antibiotic sensitivities determined. It is clear, however, that the use of antibiotics should be regarded only as primary treatment and that any predisposing factor must be dealt with, if necessary, by a surgical approach.

In patients in whom ill health has led to lowered resistance, ascending infection of any of the salivary glands may occur from the mouth via the ducts. This, once a common sequel to surgery, especially in older patients, is now not so often seen. In these cases the treatment in evidently non-surgical, being dependent on antibiotic therapy, the maintenance of a correct fluid balance, and, if possible, the resolution of the predisposing condition.

Chronic sialadenitis either of the parotid or the submandibular glands, may follow the resolution of an acute infection or may occur without any evident primary acute phase. In these circumstances symptoms are relatively low grade with tenderness and a minor degree of swelling of the affected gland, and occasionally with some degree of swelling and redness of the duct and the papilla. As in acute sialadenitis, purulent saliva may be expressed from the duct. Frequently in these cases, minor dilatations of the ductal system of the glands may be detected on sialography and, presumably, these provide loci of infection and stagnation. Such a recurrent chronic sialadenitis may occur following radiotherapy affecting the glands or may follow the minor damage due to the presence of a calculus. The principles of treatment are exactly as those of acute sialadenitis alhough in the patient with repeatedly recurrent sialadenitis of a submandibular gland, excision of the gland might be considered a reasonable form of treatment.

## Sialosis

Sialosis implies a non-inflammatory recurrent swelling of the salivary glands. There are many precipitating factors for this condition including the effect of antirheumatic drugs, drugs containing iodine, and adrenergic drugs. Similar experimental sialosis may be stimulated in animal models by the administration of isoprenalin, presumably the result of interference with the neurological stimuli controlling the gland function. On the whole these drug-induced enlargements are reversible although, in some cases, acute sialadenitis has been reported following withdrawal of the drug. Sialosis may also occur in hormonal abnormalities and in nutritional deficiency states, including anorexia nervosa and chronic alcoholism. The mechanism involved is not understood, but, histologically, there is some proliferation of serous secretory units together with atrophic changes in the duct systems.

## Necrotizing sialometaplasia

This strange condition has only recently been fully described, but there are increasing numbers of reports of cases. It appears to be some kind of vasculitic phenomenon in which ischaemic and consequential changes occur in a group of minor salivary glands, usually in the palate. The consequence is ulceration of rapid onset associated with induration which may mimic malignancy. Histologically also, the squamous metaplasia found in the salivary ducts together with pseudoepitheliomatous hyperplasia of the surrounding palatal epithelium may give a quite incorrect impression of malignancy. In fact, this is a self-limiting condition which resolves in a few weeks time without any other than symptomatic treatment. Anaesthesia of the palatal nerves has recently been reported as an early indicator of this form of ulceration. As yet there is no real information about the immediate cause of this condition or of its implications in relation to systemic factors.

## Sjögren's syndrome

Sjögren's syndrome is a condition in which three factors are associated. These are xerostomia, lack of tear secretion leading to conjunctivitis, and an autoimmune disease, most commonly rheumatoid arthritis. In the absence of the autoimmune disease the condition is often referred to as the Sicca syndrome.

Most patients are middle-aged females with a known history of rheumatoid arthritis. This is usually followed by complaints of dryness and soreness of the mouth and of the eyes. In these patients the oral mucosa tends to be atrophic and reddened, but the most frequent and most obvious change is in the tongue in which there may be painful fissures. Swelling of the salivary glands is reported in some patients, but this is not a constant finding.

It is evident that the complaint of a sore and reddened oral mucosa may be dependent upon a number of factors, and in particular those associated with haematological abnormalities. These must be eliminated before a diagnosis of Sjögren's syndrome is made. The diagnostic technique (as described by Mason and Chisholm) consists of clinical examination, parotid flow-rate estimation, sialography of a representative gland, salivary scintiscanning using labelled pertechnitate, and labial gland biopsy. The value of a salivary flow-rate estimation in establishing the true nature of the xerostomia is evident. Sialography consistently shows at least some degree of sialectisis by way of dilatation of the ducts. Pertechnitate scanning is used to assess salivary gland function—the modern computer-assisted quantitative procedures being much more helpful than the earlier qualitative methods. In the currently used method an intravenous injection of the radioisotope is followed by scanning of the salivary glands at 30-second intervals. The rate of concentration of the isotope in the glands is plotted and, after 30 minutes, salivary secretory activity is stimulated by dropping citric or tartaric acid solutions onto the tongue. The resulting activity of the glands is again followed by 30-second interval scanning. Time-activity profiles of the glands are thereby produced, and may be combined with sialographic and other studies of the glands to build up a full picture of salivary secretory activity. Finally, labial biopsy is used to demonstrate the characteristic histological changes which are evident in the small mucosal glands of the lower lip as in the major glands. These changes consist largely of the presence of many lymphocytic foci within the glands. A grading scheme has been proposed by Chisholm and Mason depending largely on the extent of this lymphocytic component.

In patients with Sjögren's syndrome there may be abnormalities of the immune system quite distinct from those associated with the rheumatoid arthritis or other autoimmune conditions. Autoantibodies may be detected, both specific and non-specific. Autoantibodies to salivary duct cells may be demonstrated in some patients, but this is felt to be associated with the rheumatoid arthritis rather than with Sjögren's syndrome itself, largely because of its erratic appearance. On the whole it should perhaps be concluded that the immunological aspects of Sjögren's syndrome are not, as yet, fully understood.

Treatment of Sjögren's syndrome is not satisfactory. It is evident that any generalized abnormality should receive suitable medical attention. The oral symptoms, essentially of dryness, can be treated only by the use of salivary substitutes such as methylcellulose mouthwashes. There is no method known of stimulating the affected glands to further action. It is probably true that, from the oral point of view, the diagnosis of Sjögren's syndrome does little to help the patient.

In Sjögren's syndrome, as in virtually all chronic inflammatory or degenerative conditions of the salivary glands whatever their basic cause, calculus formation may occur as a secondary feature, presumably as a result of stasis. Often in these circumstances the calculi may be small and multiple. Following calculus formation there may be dilatation of the ductal system of the gland proximal to the obstruction and, often, secondary infection which may be chronic and recurrent or may be acute. It will be appreciated that it is often difficult to precisely define the origin of salivary gland problems.

Treatment of this situation is often unsatisfactory; multiple courses of antibiotics may be needed in the case of recurrent infections. If it is clear that only one gland is involved excision may be considered but, by the nature of the problem, all four major salivary glands may be to some extent affected. In these circumstances it is evident that there must be a very cautious approach to surgery.

## Facial swellings

Although dental infections are by far the most common cause of swellings about the face seen by the dental surgeon, there are many other possibilities which must be considered in a differential diagnosis. Several of these possibilities have already been considered, including the swelling due to allergy or angio-oedema (often misdiagnosed as a purely dental problem) and the swellings in oral Crohn's disease or orofacial granulomatosis which are equally susceptible to a mistaken initial diagnosis. Similarly, facial swelling due to infections of nonodontogenic origin—such as, for instance, in acute sialadenitis—are commonly assumed to be of dental origin.

## Neoplasia

Neoplastic growth within the soft tissues of the face is rare. When it does occur, however, it is vital that a diagnosis should be made quickly. The signs and symptoms are of a swelling, usually growing slowly, generally painless, and not tender to pressure. Palpation reveals its solid nature. The diagnosis is entirely by biopsy examination of the tissue. Radiographic examination reveals no abnormality in an early case. If, however, there has been bone infiltration, extreme care must be taken to differentiate between this and inflammatory changes of an odontogenic origin.

## Masseteric hypertrophy

Unilateral enlargement of the muscles of mastication, and of the masseter in particular, occasionally occurs as a response to serious derangement of the occlusion leading to unilateral mastication. It may also, however, occur with little or no occlusal disharmony and, in fact, may be bilateral rather than unilateral. In the affected patients the complaint is usually only of increasing facial assymetry. On examination the masseter (or masseters) are found to be enlarged as a whole, often with a marked increase overlying the mandibular insertion. When the masseter is defined for examination by asking the patient to clench the teeth the muscle is easily palpated, the lower part often standing out and resembling a soft tissue mass. On radiography the mandibular insertion of the masseter may be represented by a marked concavity at the periphery of which there is lipping of the bone. If electromyographic facilities are available it is possible to demonstrate atypical muscle activity in all the muscles of mastication.

This is a rather mysterious condition and little is known of its aetiology. Diagnosis is entirely by clinical examination as outlined above.

## Enlargement of the lymph nodes of the face and neck

Normal lymph nodes are not palpable. Enlargement of a node or a change in texture, so that it may be palpated through the skin, indicates either that the node is taking part in the bodily defence mechanism acting against an infection or that there is some basic pathological change within the node itself. Differentiation between these two situations is vital.

Lymph nodes may be examined by palpation or by biopsy, usually of the excisional type, although aspiration biopsy may occasionally be used. When examining the lymph nodes of the head and neck each of the main groups must be palpated in turn, whilst the surrounding tissues are relaxed by bending the patient's head forward and laterally towards the side examined. From a position behind the patient the facial, submandibular, submental, parotid, cervical, auricular, and occipital groups of nodes are palpated in turn on each side. If a palpable node is found, its texture is noted and it is moved between two fingers to discover any attachment to skin, underlying tissue or adjacent nodes.

## Acute infections

Lymphadenitis arising from an acute infection such as a periapical abscess or a pericoronitis is usually unilateral. The appearance of the nodes is rapid, they are soft and are painful when touched. There may be oedema of the soft tissue surrounding the nodes giving the visual impression of greater enlargement than is, in fact, the case. The facial lymph node, lying just anterior to the anterior border of masseter at the level of the occlusal plane, is commonly involved in children and may give rise to difficulty in identification.

## Chronic infections

In chronic infections the affected nodes are firm and not tender. They do not become attached to the skin or surrounding tissues and do not coalesce. In long-standing cases of chronic infection, calcification of a node may present as a solitary hard, non-fixed swelling. More widespread smaller calcifications in the cervical nodes are more common. They are usually discovered incidentally during radiography, are asymptomatic and cannot be palpated.

Widespread lymphnode enlargement may be the first clinical sign of the HIV virus—the submandibular nodes are often prominently affected. The presence of painless enlarged lymph nodes, particularly in the superclavicular area (the 'syphilitic collar') is a classic marker of syphilis in its secondary and later stages.

## Secondary neoplasia

Spread to lymph nodes of the neck may occur at any stage in the growth of a malignant neoplasm of the oral cavity, pharynx, antrum, or related structures. The nodes are initially painless, hard, and usually unilateral. In the early stages the enlarged lymph nodes may show only non-specific inflammatory changes on biopsy. In later stages the nodes may become painful because of secondary infection, may coalesce, or may become fixed to the skin or surrounding tissues. Diagnosis is confirmed by excision biopsy of the nodes.

# 6

# Lesions of the teeth and periodontium

THE teeth are poor indicators of generalized disease. Following their calcification metabolic processes have little effect on the structure of the teeth and the only major changes which ordinarily occur are the loss of substance caused by caries, attrition, erosion, and abrasion. Other structural abnormalities almost invariably reflect changes occurring in the period during which the teeth were being formed. Apart from structural variation, abnormalities in the numbers, size, and shape of the teeth occasionally occur in conjunction with abnormalities of the bones or of the skin and other epidermally derived structures. When numbers of missing teeth, supernumerary teeth or abnormally shaped teeth are observed it is as well to consider the possibility of some such complex association. It must always be remembered that teeth missing from the arch may be either congenitally absent, have been extracted, or may be unerupted.

## Partial anodontia (hypodontia)

Partial anodontia represented by the loss of one or two teeth with no apparent associated abnormalities is not uncommon. World wide studies show that the teeth thus missing vary from population to population. A study carried out in an English population and excluding the third molars has shown that the teeth most likely to be missing are the lower second premolars (40.9 per cent) followed by the upper lateral incisors (23.5 per cent) and by the upper second premolars (20.9 per cent). Other teeth are much less frequently missing. in the case of the upper lateral incisor it has been shown in several populations that the absence of this tooth (or teeth: the loss is often bilateral) is genetically determined. The third molars also show a wide variation in the pattern of absence determined in differing populations. Most surveys, however, show that one or more third molars are missing in approximately one-quarter of the population.

A characteristic finding in partial anodontia is the presence of small and conically shaped teeth replacing normal units of the dentition (Fig. 6.1). These simple structures represent a single cusp rather than the fused series of cusps which together form the normal tooth. It appears that these simple teeth represent a half-way stage to total suppression. It is not uncommon, for example,

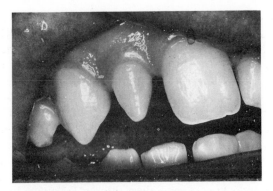

Fig. 6.1. A conically-shaped tooth in the position of an upper lateral incisor.

for one upper lateral incisor to be completely lost whilst the opposite tooth is found to be of the conical type.

There has been little work carried out on partial anodontia in the primary dentition, but it would seem that this is a relatively rare occurrence. It does seem, however, that from time to time the same tooth may be missing from the deciduous and from the permanent dentition.

Partial anodontia associated with abnormalities of the bone or ectodermal appendages is relatively rare. The dysplasia involved may be attributed to ectodermally derived structures (as in anhydrotic ectodermal dysplasia Fig. 6.2) or with more complex syndromes in which there are both dermal and bony abnormalities.

Supernumerary teeth may occur at almost any point in the dental arch and are usually of little other significance. It must be remembered, however, that a

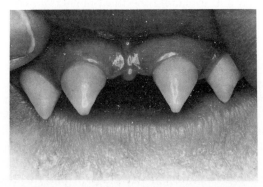

Fig. 6.2. Multiple conical teeth associated with partial anodontia (hypodontia) in anhydriotic ectodermal dysplasia.

frequent consequence of the presence of the supernumerary teeth in the anterior region of the maxilla is a failure of the normal teeth to erupt. Thus, the most common sign of the presence of such supernumerary teeth is the absence of a permanent incisor from the arch. Patients with cleft palates are fairly frequently found to have supernumerary teeth. These are most often of the small conical type and closely associated with the margins of the cleft.

The only more widespread condition in which the presence of many supernumerary teeth is common is cleido-cranial dysostosis. In this abnormality of membrane bone formation the changes observed are lack of calcification of the clavicle, flattening of the frontal bone, and the presence of a number of supernumerary teeth. These teeth are often of complex form, resembling units of the normal dentition, and frequently remain unerupted (Fig. 6.3). A parodoxical feature of this condition is that it may appear that the patient is suffering from partial anodontia because of the failure to erupt of large numbers of teeth within the jaws. It is often quite impossible to distinguish the teeth of the normal dentition from those of the abnormal dentition. The presence of such numbers of unerupted teeth of mature form is quite characteristic of cleido-cranial dysostosis and should be followed up by the taking of confirmatory radiographs of the skull and clavicles.

## Variation in dates of eruption of the teeth

The wide normal range makes it difficult to specify accurately the dates of eruption of either the deciduous or the permanent teeth. A number of factors has been found to affect the date of eruption, including racial origin and such unlikely influences as the socio-economic environment. In general, earlier bodily development is reflected in early eruption of the teeth.

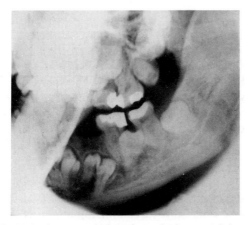

Fig. 6.3. Multiple unerupted teeth in cleido-cranial dysostosis.

Markedly premature eruption of the permanent teeth is very rare. It has been suggested that it may occur in cases of hypersecretion of those hormones which influence development. It is somewhat more common to note premature eruption of the deciduous teeth. Frequently, no systemic factor is found to account for this. It should perhaps be mentioned that teeth present at birth in a few children (the neo-natal teeth) do not represent premature eruption. These are supernumerary teeth and part of a separate pre-deciduous dentition.

Delayed eruption of the deciduous teeth may occur in endocrine deficiency states and it has been shown that in Down's syndrome not only are the eruption dates somewhat retarded in general, but also there is often an unusual sequence of eruption. It is very difficult to ascribe a characteristic dental picture to many of the endocrine abnormalities since, in a number of cases, varying and contradictory effects have been described. In many such patients the most obvious abnormality is disproportion in the sizes of the teeth and the jaws, this in turn leading to gross irregularity of the occlusion.

## Variation in the size of teeth

The size of the teeth of any individual is determined largely by inherited factors. Extreme variation in size, either in the direction of small teeth (microdontism) or in the direction of large teeth (macrodontism) may be accompanied by no other growth abnormality. Conversely, endocrine growth disorders leading to gigantism or dwarfism may be accompanied by no corresponding variation in tooth size. A few cases of hemifacial gigantism have, however, been recorded in which unilateral macrodontia has occurred.

It follows that the size of the teeth alone bears no relationship to metabolic factors in the vast majority of instances and that diagnostic significance cannot be given to such variation, excepting as an observed developmental factor.

## Variation in cusp and root form of teeth

Variation in cusp form of the teeth usually extends to the suppression or exaggeration of the normally present cusps, although extra cusps may be present. These extra cusps probably represent an adventitious extra fold in the enamel organ with no other significance. Similarly, the presence of extra roots has little other than local significance, although the presence of bilateral extra roots, as frequently occurs, suggests rather more than a random accidental splitting of Hertwig's sheath during root formation.

## Erosion

Erosion is the loss of dental hard tissue by a chemical process which does not involve bacteria. The active substances may be acid spray (as in the metal finishing industries), medicines of low pH, or acidic foodstuffs and beverages. These include citrus fruit and juice, and many soft drinks including fruit squashes and cordials, Mixer type drinks, and many others. A highly significant

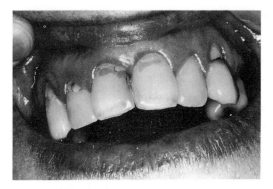

**Fig. 6.4.** Acid erosion with cervical lesions and loss of normal enamel contour.

cause of erosion is the self-induced vomiting of bulimia nervosa which may result in widespread loss of enamel, particularly from the palatal surfaces of the upper anterior teeth. As is well known, young females are by far the most commonly affected, but a few male patients have been described (Fig. 6.4).

## Discolouration of the teeth

Widespread colouration of the teeth occurs in a few diseases in which abnormal blood pigments circulate. Of these infantile jaundice is the most common. In this condition the deciduous teeth may be coloured blue-green due to the laying down of a pigment in the immediate postnatal dentine zone. A less common and now virtually eliminated cause of tooth discolouration is haemolytic anaemia of the newborn caused by Rh incompatibility. Following the haemolysis pigments may be deposited in the skin and in the teeth which may take on colouration which varies from grey to green/grey and to brown. Colouration of the teeth also occurs in some other considerably rarer situations in which abnormal pigments circulate, for example, in porphyria. However, a far more common cause of tooth discolouration is tetracycline staining.

It need hardly be said, however, that the number of young patients with tetracycline staining is rapidly reducing with the almost total withdrawal of this group of antibiotics from at risk groups.

## Disturbances of enamel and dentine formation

When the normal sequence of enamel matrix formation and calcification is disturbed a series of abnormalities may be produced. These may be distinguished as hypoplasia, when the quantity of enamel is reduced, or as hypocalcification in which the degree of calcification is unsatisfactory. The two conditions may be combined and various clinical conditions may be differentiated. A parallel range of disturbances in the formation of dentine may also occur, but these are not so well differentiated as in the case of enamel.

The use of the term 'hypoplasia' requires some explanation, since it is used both in the strictly scientific sense as mentioned above and also in a clinical sense to describe a generalized disturbance of enamel structure caused by some form of systemic disturbance.

This group of conditions can be conveniently divided as follows:

(1) enamel hypoplasia resulting from a local or systemic disturbance during its formation;

(2) genetically determined defects of enamel formation;

(3) genetically determined defects of dentine formation.

## Hypoplasia due to local infections

When an infection occurs in association with a deciduous tooth the permanent successional tooth developing below it may undergo a disturbance of development. In such cases the enamel is usually distorted and pitted (Fig. 6.5). This condition is easily recognized by its restriction to a single successional tooth.

## Hypoplasia due to generalized infections and trophic disturbances

Widespread infections or other trophic disturbances during the tooth enamel development period may adversely affect the laying down of enamel. Such conditions affect all the teeth developing at the time and it is often possible to time accurately the onset of the disturbance by the position of the defect on the teeth. In general, the deficient enamel forms a band around the tooth corresponding to the period of disturbance (Fig. 6.6).

The band may be wide or narrow and, in some circumstances, the banding

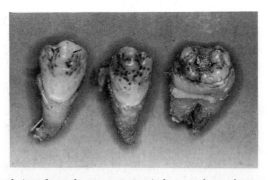

**Fig. 6.5.** Three teeth (not from the same patient) showing hypoplastic enamel, the result of infection associated with a deciduous tooth.

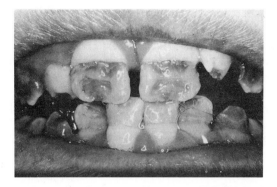

**Fig. 6.6.** Hypoplasia as a result of a generalized trophic disturbance during development.

may be incomplete. It has been suggested that the enamel opacities seen in some patients may represent a minor form of this condition.

Histologically, the dentine developed at the same time as the affected enamel may show slight deficiencies, but this does not result in clinical abnormality.

Unless the disturbance of tooth formation is unusually severe the teeth do not seem to be unduly susceptible to attack by caries.

Prenatal syphilis may produce defects in tooth development since the spiro-chaete may lodge in the enamel organ and interfere directly with the formation of the enamel. In general, this occurs after the differentiation of the deciduous teeth and before the differentiation of the premolars. Thus, the effects are generally confined to the anterior permanent teeth and to the first permanent molar. The typical molar form of the syphilitic tooth is the mulberry molar in which the shape of the tooth is well expressed by its name. The typical variation in the anterior teeth takes the form of a conical shape and notched incisal edge—Hutchinson's incisor (Fig. 6.7).

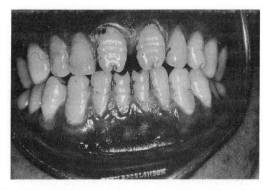

**Fig. 6.7.** Hutchinson's incisors in congenital syphilis.

## Hypoplasia due to fluorosis

If the development of the teeth occurs when large amounts of fluorides are ingested mottling of the enamel may occur. The effect of fluorosis may be recognised by the presence of opaque white patches in the enamel, often arranged in a band like formation. Unlike the teeth in other forms of hypoplasia the teeth affected by fluorosis are susceptible to a brown discolouration of unknown origin which may somewhat confuse the diagnosis (Fig. 6.8). Similar idiopathic mottling may occur in teeth of patients from non-fluoride areas, but this is rare.

## Amelogenesis imperfecta

Amelogenesis imperfecta is a hereditary developmental defect of enamel. The condition may show as either a hypoplasia of the enamel or as a hypocalcification. This depends on the stage of enamel formation which is disturbed in the condition. If an early phase of enamel formation is disturbed the amount of matrix laid down is reduced, but the calcification is complete. We thus have a thin and irregular layer of hard enamel. This is the hypoplastic type. If a later stage of enamel formation is disturbed we have a normally thick layer of poorly calcified enamel. This is the hypocalcified type.

In the case of hypoplasia the enamel is seen to be deficient, but hard over the whole of the dentition (Fig. 6.9). Severe attrition may occur early in life. In the case of hypocalcification the whole of the enamel is soft and eroded with loss of much enamel by attrition and exposure of the dentine (Fig. 6.10). In both forms the deciduous and permanent dentitions may be affected. The dentine retains its normal structure in both cases.

## Dentinogenesis imperfecta

In this condition there is failure of development of the dentine with normal

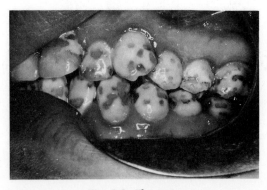

**Fig. 6.8.** Fluorosis.

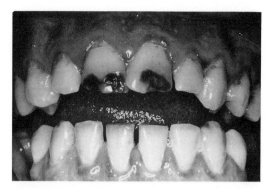

Fig. 6.9. Hypoplastic enamel. The enamel is imperfect, but is hard.

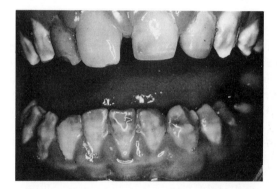

Fig. 6.10. Hypocalcified enamel. The enamel is soft and eroded.

enamel development. It is a hereditary condition and, like amelogenesis imperfecta, occurs in the deciduous and permanent dentitions. The teeth are usually of normal morphology, but grey or brown in colour (Fig. 6.11). They show a rather irridescent colouration which leads to the term hereditary opalescent dentine. The pulp chambers of affected teeth are often reduced in size compared to the normal and may, in fact, be obliterated. Although the enamel is of normal structure it readily breaks away, leaving the dentine exposed.

Occasionally, this condition occurs as part of a generalized condition of osteogenesis imperfecta in which imperfect calcification of the bones leads to frequent fractures. Often in such cases there is a deficiency in the sclera of the eye leading to a blue colouration. This does not occur in simple cases of dentinogenesis imperfecta.

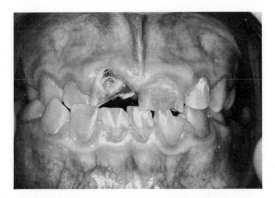

Fig. 6.11. Dentinogenesis imperfecta.

## Hypoplasia associated with generalized disease

Hypoplasia or hypocalcification of the enamel may occur in patients suffering from generalized ectodermal disease (such as epidermolysis bullosa) or from disturbances of calcium metabolism such as hypoparathyroidism. These conditions clearly represent the final result of abnormal tooth germ formation and abnormal calcification, respectively. Many other diseases may produce dental abnormalities of a similar type. These changes are, however, rare. It is unlikely that the generalized disease process will have gone unrecognized by the time that the tooth abnormality has become evident in the majority of these patients (Fig. 6.12).

It will be evident that the diagnosis of the abnormalities of tooth structure described above depends largely on the recognition of the clinical appearance of

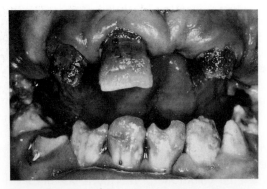

Fig. 6.12. Hypoplasia in a patient with disturbed calcium metabolism.

the teeth. Radiography apart, there is very little in the way of supplementary tests or investigations which will add to a carefully carried out clinical examination of the patient, an examination which should evidently include a careful medical and family history. In virtually all cases laboratory tests prove unproductive.

## The periodontium

Although gingival changes may occur in some generalized conditions the gingivae are usually not particularly good diagnostic indicators of the disease process involved; the changes are often clinically non-specific. The histological changes in abnormal gingivae are also often very difficult to interpret. The pre-existing inflammatory cell infiltrate and, often, secondary inflammatory features present in the gingivae confuses the picture, and make gingival biopsies much less useful than might be expected. If the alternative exists between taking a gingival biopsy and one from another area; for example, the buccal mucosa, the gingival site should almost always be avoided.

A few systemic conditions have been described in which clinically recognizable gingival changes occur; for instance, in Wegeners Granulomatosis, an uncommon vasculitic disease with widespread systemic implications, a quite characteristic gingival picture has been described although it certainly does not occur in all cases (Fig. 6.13). Such specific and recognizable changes are rare compared to the non-specific gingival changes which may occur in other disease processes and which represent an accentuation or modification of the widespread changes of chronic gingivitis which occur in a large proportion of the population. Thus, the initial gingival changes in diabetes is a simple hyperplastic gingivitis. In other circumstances, for instance, in the skin diseases lichen planus and pemphigoid, the gingival changes are the result of a collision lesion between the generalized condition and pre-existing or superimposed changes of chronic

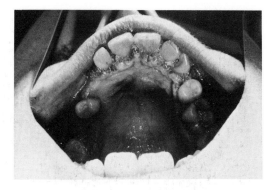

**Fig. 6.13.** A specific gingivitis—that in Wegeners granulomatosis.

gingivitis. The non-specific nature of the resulting gingival changes is illustrated by the adoption of the term 'desquamative gingivitis' which is used to describe the pattern reaction of the gingivae in both these (and probably some other) conditions. There has been, in the past, a tendency to attribute desquamative gingivitis in female patients to undefined hormonal disturbances, but this is no longer accepted.

Gingival lesions in diseases of the skin are discussed in Chapter 8, endocrine-induced changes in Chapter 9.

It is a natural consequence of gingival abnormality that plaque accumulation and false pocketing may lead to more profound periodontal changes. An example of a condition in which this may happen is gingival fibromatosis, a genetically determined condition in which there is non-inflammatory hyperplasia of the fibrous component of the gingivae with consequent gross distortion of normal gingival architecture (Fig. 6.14). The build-up of plaque may be difficult to avoid, with consequential secondary effects such as calculus formation, the extension of pocketing, and loss of periodontal attachment. However, there are some genetically determined conditions in which there does appear to be a predisposition to rapidly progressive periodontitis as a primary feature. Patients with Down's syndrome are in this group. Another more uncommon condition is the Papillon–Lefevre syndrome, in which tylosis (hyperkeratosis of the palms and soles) is associated with rapidly progressive periodontal disease which may affect both the deciduous and the permanent dentitions. This is a quite different association of tylosis with an oral condition than that mentioned in Chapter 7 (tylosis, oral leukoplakia, oesophageal carcinoma).

The relationship of immune deficiency and iatrogenic immune suppression to periodontal disease is an interesting one; there are quite conflicting accounts of the effects of both these on the periodontal structures. It does seem, however, that in patients with primary immunodeficiencies, gingivitis is less marked than in comparative immunocompetent groups. Presumably, this is the result of

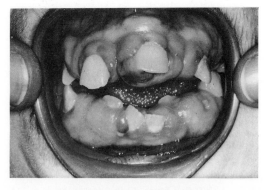

**Fig. 6.14.** Gingival fibromatosis.

natural supression of the immune mechanisms thought to be responsible for at least part of the damage in periodontal disease. Similarly, there are reports that highly immunosupressed renal transplant patients taking steroids and steroid sparing drugs show less gingival inflammation than comparative control groups. This is in spite of the fact that such patients may develop a wide range of oral infective lesions (Chapter 9). Many renal transplant patients are now maintained on other regimes including cyclosporin A or nifepidine, both of which may cause a marked hyperplastic gingivitis resembling that induced by phenytoin (Chapter 9). In a few AIDS patients a rapidly progressive periodontitis is part of the picture; the mechanism of this has not yet been elucidated.

Gingival lesions are well recognized in leukaemias (both early and late) and in abnormalities of neutrophil numbers and function. These are discussed further as are nutritional factors in Chapter 9.

# 7

# White patches: overgrowths and neoplasms

## Leukoplakia

Although many lesions of the oral mucosa may appear in the form of white patches, the term leukoplakia (white patch) is specifically used to describe those lesions in which the basic aetiology is a disturbance of the normal maturation processes of the epithelium, a disturbance which often leads, in its turn, to changes in the surface layer of keratin. The change in the pattern of keratinization is, however, a relatively insignificant factor compared to the importance of the underlying cellular abnormalities on which the subsequent behaviour of the lesion depends. Leukoplakia is a term which, in the past, has been employed in a number of different ways: in particular, it has been widely used to imply a premalignant condition. The assessment of premalignancy in these lesions is not, in fact, an easy matter and it is not possible to judge premalignant potential on clinical grounds alone with any degree of accuracy. In order to clarify the position a definition has been proposed by Pindborg and this is now generally accepted as a reasonable definition of leukoplakia, it has been adopted, for instance, by the World Health Organization for use in epidemiological surveys. This definition is of a white patch on the oral mucosa which cannot be wiped away and is not susceptible to any other clinical diagnosis. By this definition superficial plaques which can be wiped off (such as those of thrush) are excluded, as are the white lesions which affect the oral mucosa in a number of diseases of the skin, these latter being clinically recognizable. Leukoplakia may be described as a lesion of non-specific histology with a variable behaviour pattern and with an unpredictable (but statistically assessable) tendency to malignant transformation.

In the study of leukoplakia, it is necessary to consider a number of closely related lesions which, even though they may not fit the precise definition of being white patches, are similar in histological and clinical detail to leukoplakia and, therefore, must be considered as variants of that condition. Of these, preleukoplakia most resembles leukoplakia itself, and presents as a rather diffuse grey lesion which is considered to be a halfway stage to the production of leukoplakia proper. It is not by any means clear whether all pre-leukoplakias progress to become leukoplakias, but there is no doubt that in some cases this transformation does occur. Erythroplakia, as its name implies, is essentially a red lesion of the oral mucosa. It has a smooth velvety appearance, but there may be many small areas of leukoplakia superimposed on the red background, the lesion then being referred to as speckled leukoplakia. Although it has been made clear by many clinical-histopathological surveys that it is impossible to make an assessment of the eventual behaviour of leukoplakia without histological study, there are generally accepted patterns of behaviour associated with the clinical variants which give at least some indications as to the likely eventual prognosis. For example, there is no doubt that both erythroplakias and speckled leukoplakias are more prone to undergo malignant change than are simple leukoplakias. However, statements of this nature are generalizations and cannot be applied to any specific lesion with certainty.

## Aetiological factors

As has been pointed out in Chapter 1, the integrity of the epithelium depends on a number of factors. In the first instance, there are the genetic instructions contained within each basal cell as it is produced by mitosis. It is a matter of some argument as to whether the genetic information within each epithelial cell is so specific as to determine exactly whether the cell is eventually to become, for instance, a fully keratinized cell or be incompletely keratinized. However, there is some evidence that, whatever the precise nature of the genetic coding, the eventual fate of each epithelial cell is decided at the time of division. Thus, any change in the structure of the epithelium is likely to occur from the basal layer and not from a layer above it. It seems very likely that, in certain conditions, changes in the epithelium are involved with changes in the underlying connective tissues and, in some of these, it seems probable that the initial lesion lies within the connective tissue whereas the visible and, perhaps more significant part of the lesion, lies within the epithelium. A further factor which is known to affect the structuring of the epithelium is the application of various forms of chemical irritation. By far the most important from the clinical viewpoint are the substances applied to the oral mucosa as a result of tobacco smoking and chewing habits. Yet another factor in the production of epithelial abnormalities is that of infection. There is no doubt that a close association exists in some cases between infection by *Candida albicans* and the presence of leukoplakia. Moreover, although the matter is far from proved, there have been suggestions (largely based on immunological studies) that there may also be a possible participation of the herpes virus in the production of oral leukoplakia. Perhaps the most obvious form of extrinsic stimulus leading to epithelial abnormality is mechanical irritation. This will be discussed below.

Apart from abnormalities induced by external factors there are probably many random variations from the normal structure of epithelial cells which are monitored by the immune defence mechanism. There is considerable evidence that the cell-mediated immune system in particular is involved in maintaining the integrity of the epithelium (and other tissues) in this way, and it is evident that deficiencies or suppression of the immune system will lead to an increased incidence of abnormality. This situation is of clinical significance in that heavily immunosuppressed patients have been found to exhibit a far higher incidence of neoplasia than the population as a whole. To summarize then, it is evident that the maintenance of the integrity of the epithelium depends on a balance of factors. On the one hand, there are the essential stabilizing factors which control the rate of mitosis and ensure normal structuring, and on the other hand, there are the unstabilizing influences of applied irritants and mesodermal tissue changes. Should abnormalities occur it presumably depends on the integrity of the immune system as to whether the abnormal cells produced are recognized and rejected as 'non-self' at mitosis, and if applied or intrinsic irritants or unstabilizing factors co-exist with a deficient immune defence mechanism then lesions will be produced. In such a patient with an inefficient surveillance mechanism, application of any of the possible aetiological agents may result in the production of a lesion.

A further, but poorly understood, predisposing factor in the production of white patches of the oral mucosa is the effect of mechanical trauma. It is well known that such trauma may have a wide variety of effects on the mucosa, depending on the exact nature of the trauma, its rate of application and (presumably) the individual response. When the trauma is acute and localized the epithelium is destroyed and an ulcer is produced. If, however, the trauma is less acute and less localized then there will be a range of responses depending on the precise clinical conditions. If the irritation is relatively chronic in character and of low intensity, the major change induced in the oral mucosa may well be an abnormality in the keratinization pattern (Fig. 7.1) which must be associated

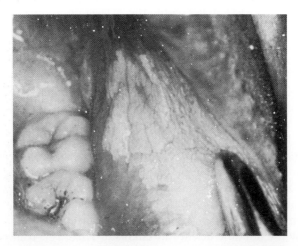

Fig. 7.1. Traumatic keratosis of buccal mucosa caused by friction from denture.

with some degree of change in the underlying cells. There is no doubt that such traumatic keratoses exist and that the large majority are reversible, completely disappearing if the trauma is removed. Nevertheless, what is not so clear is the possible relationship between such traumatically induced conditions and their eventual transformation to irreversible lesions with the clinical characteristics of leukoplakia or even carcinoma. The statistics relating to such changes are not known, but it is clear that such transformations occur relatively infrequently. It is of interest to recall that the mucosal response to trauma most often seen in the oral cavity is that of the denture granuloma; this exhibits an extremely low rate of malignant transformation. The principal change in this particular lesion is, of course, in the connective tissue component of the mucosa rather than in the epithelium.

Leukoplakia appears as an intrinsic white area of the oral mucosa, sometimes homogeneous, sometimes wrinkled and sometimes with a verrucous or fissured surface (Figs 7.2–7.4). The density of the white patch may vary from a transparent filmy appearance to a dense thick one. In some cases the leukoplakia is a single discrete and well-bounded area, whilst in other patients there may be widespread abnormality of the mucosa with a number of lesions distributed in varying sites. In a Danish population the highest incidence of leukoplakia has been found to be on the buccal mucosa and commissures, and then, in descending order of frequency, on the alveolar ridges, tongue, buccal sulci, floor of mouth, labial mucosa, and palate. However, in common with all other statistics regarding leukoplakia and associated lesions, it must be remembered that the figures refer to a specific population and that other populations in different areas and with different influencing factors (such as tobacco habits) may show diverse clinical features.

Two groups of leukoplakias may be recognized: idiopathic leukoplakias in which no aetiological factors have been recognized, and leukoplakias in which

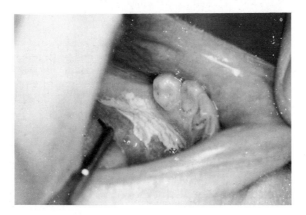

**Fig. 7.2.** Leukoplakia of floor of mouth. This shows the so-called 'ebbing tide' pattern.

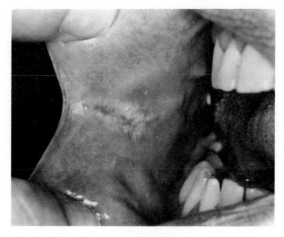

Fig. 7.3. Leukoplakia of the commisure and buccal mucosa.

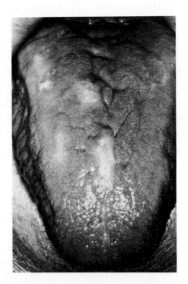

Fig. 7.4. Leukoplakia of the tongue.

an evident predisposing factor may be present. It is probable that the ideopathic group may include a certain number of lesions in which the aetiological factor remains unrecognized, but even with this reservation, it would seem reasonable to accept the possibility of the existence of idiopathic lesions. The term 'senile keratosis' has been used to describe idiopathic leukoplakia arising in older patients, but there is no evidence to suggest the existence of a different aetiology in this group. It would seem likely that the lesions represent exactly the same abnormalities as those found in younger subjects; with advancing age it is more likely that abnormalities will develop because of the longer period involved.

It is clearly established that there are a number of clinically significant aetiological factors which may contribute to the production of oral leukoplakia. The most important of these is the use of tobacco, either when smoked or chewed in one or other of the large number of tobacco-using habits which have been described. The precise nature of the action of tobacco or of its smoke on the oral mucosa is not yet known, but there can be no doubt that a profound effect is often exerted. Apart from the tobacco itself other substances involved in tobacco chewing habits (such as betel-nut and lime) may also be implicated in the production of leukoplakia. There have been many surveys concerned with the effect of tobacco habits on the oral mucosa and, in virtually all cases, it would appear that those subjects who use tobacco in any of its forms are much more likely to develop oral leukoplakia. It is also well established that the precise nature of the tobacco habit is of great significance in the determination of the exact form of the lesion produced and also of its eventual prognosis. In an investigation of the relationship between the incidence of leukoplakia and the sex, age, and tobacco habits of a Bombay population, it was shown quite clearly that the most important extrinsic factor by far was the smoking habit and that the increased incidence of lesions in males was directly attributable to the increased use of tobacco in males as against females. It is not difficult to extrapolate these findings to other series in which there has been a constant marked predominance of leukoplakia in male subjects. There is no doubt that those leukoplakias which are directly attributable to tobacco habits are, to a large degree, reversible and it has been clearly demonstrated on many occasions that regression of a lesion may be induced by the cessation of the habit, although the regression may not be permanent.

A second important aetiological factor in the production of oral leukoplakia is that of infection and, of the organisms involved, it seems that *Candida albicans* is by far the most important. It is well established (as has been described in Chapter 3) that there may be a heavy infiltration of the pseudohyphal form of *Candida* into the epithelium of oral leukoplakias and it has been shown that such lesions (*Candida* leukoplakias) are associated with marked epithelial irregularities and an increased incidence of malignant transformation. The precise relationship between the candidal infection and the production of the leukoplakia is not known and, in particular, it has not yet been determined whether the infection is a primary or a secondary feature. Histological studies would seem to provide evidence for either viewpoint, and it will probably be as the result of immunological studies that the final decision as to the exact aetiological factor in these cases is made. However, recent work has shown that, on an experimental basis, candidal infiltration into epithelium may of itself produce changes resembling those of leukoplakia, and there would also seem to be evidence that complex immune deficiencies may occur in some patients with leukoplakia, including a deficient immune response to *Candida albicans*.

There is also some evidence of an additive effect between the actions of tobacco and *Candida* in the formation of leukoplakias. It is currently suggested that heavy smokers are at increased risk, firstly of developing *Candida* leuko- plakia and, subsequently, developing carcinoma within it. This is not a proven association but there is a considerable bulk of anecdotal information to support the concept.

It is not easy to ascribe a characteristic appearance to *Candida* leukoplakias, and, indeed, a wide variety of lesions may be shown to include the organisms. However, it is generally accepted that *Candida* leukoplakias often have a somewhat irregular and nodular appearance, whilst a high proportion of commissural lesions extending to the external angle of the lips are *Candida* leukoplakias (Fig. 7.5). Diagnosis of a *Candida* leukoplakia cannot be made by taking superficial swabs—the *Candida* are within the lesion and very few may be on the surface. For a reliable diagnosis it is necessary to stain sections from a biopsy specimen with PAS reagents; in this way the hyphae are readily seen within the superficial layers and stratum corneum of the epithelium. In view of the increased rate of malignant change in *Candida* leukoplakias it is of evident importance that these lesions should be recognized and treated with heightened suspicion.

The role of syphilis in the induction of oral leukoplakia has, in the past, been well recognized. Such lesions characteristically involved the tongue and were notorious for displaying an extremely high rate of malignant transformation. With the more successful treatment of early syphilis, however, such lesions have virtually disappeared (this is true of European conditions, but might not be so in other communities). The precise nature of the aetiological process in these cases is not clear, but it has been pointed out that syphilitic patients show a deficient cell-mediated immune response. It may well be that the consequent loss of

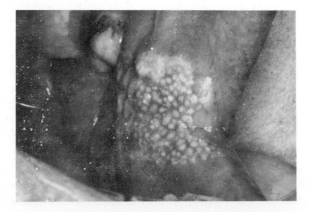

Fig. 7.5. Candidal 'speckled' leukoplakia at the commisure.

immune surveillance is the most important factor both in the origin and the subsequent malignant transformation of these leukoplakias.

The possible role of the herpes virus in the induction of oral leukoplakias has been previously mentioned, but it must be pointed out that, at the present time, the evidence of such association is speculative and by no means widely accepted. However, the relationship between viral infection and neoplastic or similar changes is becoming more clearly defined, and it may well be that, in the future, such a relationship may be proved in the case of oral leukoplakia.

The role of alcohol in causing abnormalities of the oral mucosa is doubtful. Practically all available information refers to oral carcinoma rather than to leukoplakias and it is extremely difficult to disentangle the effect of alcohol itself from the complex of other factors involved. It is probably true to say that no convincing evidence has, as yet, been produced to show that alcohol consumption has any clinically significant effect on the oral mucosa when considered in isolation from other factors such as tobacco consumption. A few animal experiments have suggested that alcohol may possibly have such an effect, but this is as a secondary effect of liver damage rather than as some form of primary irritant.

There is no doubt that some leukoplakias undergo malignant transformation, the overall incidence of such change is approximately 5 per cent. It has also been established that erythroplakias, speckled leukoplakias, and *Candida* leukoplakias all undergo a significantly higher rate of malignant change, perhaps of the order of 15–20 per cent. However, as has been previously pointed out, clinical assessment alone is valueless in determining prognosis and it is necessary to carry out histological studies in order to obtain any useful idea of possible eventual behaviour. Even this, however, is by no means a simple matter and there is a great margin of possible error in the usual methods of subjective histological assessment. Such features as abnormal mitotic activity or the presence of dyskeratosis have long been recognized as being pointers towards possible or probable malignant transformation, but the significance of other features or of combinations of abnormalities has been extremely difficult to assess. This has led to the introduction of methods of systemic assessment including a technique of scoring in which various abnormalities or 'atypia' of epithelial structure are given a weighted score. When these are added together, the total score may be taken to indicate the degree of demonstrable premalignant tendency in the lesion. The atypia involved in the assessment include abnormalities of the individual cells (such as the presence of abnormal mitotic activity or pleomorphism) and also variation in the structure of the epithelium as a whole. It is, as yet, too early to assess the accuracy of this method in predicting the eventual behaviour of the lesions studied, but there can be little doubt that this form of approach represents a marked improvement on the all round subjective assessment which has previously been accepted as the only monitor available. It must be realized, however, that whatever method of assessment is used, there are occasions when a completely false impression of benignity may be gained from the histological examination of the tissue. It is, therefore, imperative that all leukoplakias and similar lesions should be kept under long-term review, even though biopsy may reveal an apparently inno- cuous histopathological appearance.

A further factor which may alter the statistically determined prognosis of a leukoplakia is that of site. There are some factors contributing to this which are quite understandable; for instance, the high incidence of candidal leukoplakias near the commisures implies a high rate of malignant transformation in this site. However, it is by no means clear why floor of mouth leukoplakias (such as that shown in Fig. 7.2) should be recognized as having a potential for transformation considerably higher than that in the mouth as a whole. It has been suggested that this is due to the pooling of carcinogens in the floor of the mouth, but since the carcinogens in question have not been identified, this remains an interesting speculation.

Apart from this kind of histological assessment, attempts are currently being made to determine premalignant characteristics in leukoplakias by the measure- ment of various biochemical features of the epithelium. Assays of enzymatic activity in the epithelium by histochemical techniques have been carried out and the preliminary results suggest that clinically significant information may eventually be obtained in this way. Similarly, measurements of DNA synthesis within the epithelial cells have shown distinct variations in those lesions which eventually become malignant, and this is also a technique which may in the future be translated into usable clinical terms. A recently reported technique involves the assessment of surface characteristics of the cells of the lesions by scanning electron microscopy. It would seem that variations in the morphology of the cells may be correlated with the light histological appearance of the lesions as a whole. If this technique is developed into a clinically viable method it will have the great advantage that exfoliated cell specimens might be used for assessment. Currently, however, it should be emphasized that the only method of assessment of premalignant potential which is reasonably well understood and which may be credited with at least some degree of accuracy is that of the histological assessment of atypia.

There is currently great interest in the existence of cell surface markers and genetic markers (oncogenes) which are thought to indicate the inherent potential of the individuals carrying them to develop malignancy. These investigations are currently being extended to determine the possible presence of similar markers in potentially malignant lesions such as a leukoplakias as a method of predicting subsequent behaviour. The development of monoclonal antibodies (Chapter 2) has greatly accelerated work of this kind.

## Erythroplakia

As implied by its name, erythroplakia is essentially a red lesion of the mucosa. An alternative term—erythroplasia—is taken from the name given to a somewhat similar lesion of the penis (erythroplasia of Queyrat), which has for some time been recognized as a premalignant lesion. The oral lesion is relatively uncommon and appears as a bright red patch, which is well defined from the surrounding mucosa and has a velvet-like surface texture. There may be occasional white areas adjoining the lesion and, in some cases, patches of erythroplakia may be interspersed with leukoplakic lesions. It is most often seen on the buccal nucosa, although it may occur, less commonly, on other sites of the oral mucosa.

The most important feature of this lesion is the high incidence of epithelial atypia shown on histological examination, these are associated with atrophy of the epithelium. A significant number of these lesions display sufficient epithelial atypia to be assessed as carcinoma *in situ*, a diagnosis which warrants immediate prophylactic surgical intervention. On the other hand, it has been pointed out that some of these lesions display virtually no atypia and, indeed, may be of a simple inflammatory nature. It is quite evident that biopsy examination is essential in all these cases. There is some discussion as to whether the presence of *Candida* infection in these lesions is a factor which leads to a more active epithelium. As in the case of leukoplakia it is by no means clear as to whether the *Candida* is a primary or a secondary feature of the lesions. It is generally felt, however (perhaps without any clear evidence) that the presence of *Candida* in the epithelium is likely to coincide with an increased tendency to malignant change. Whatever the aetiological factors involved, there seems little doubt that a high proportion of lesions clinically diagnosed as erythroplakia present sufficient epithelial abnormalities to be classified on histological examination as incipiently malignant. It would seem that erythroplakia warrants the description which has been applied to it of being the most serious of the oral precancerous lesions.

## Speckled leukoplakia

This lesion can be considered either as a variant of leukoplakia or of erythroplakia. It appears as a series of white nodular patches on an erythematous background and so, in appearance, is midway between these two conditions. Any area of the oral mucosa may be involved, but many lesions appear on the buccal mucosa near to the commissures (Fig. 7.5). Histologically, these lesions show a high incidence of atypia and, in many cases, the presence of candidal hyphae in the epithelium. The same doubts have been expressed about the exact role of *Candida* in these lesions as in the case of homogenous leukoplakias, but there is no doubt that there are frequently two coincidental factors in all these lesions: the presence of *Candida* and the presence of epithelial atypia. Whether there is a causal effect remains to be determined. These lesions should always be considered as being potentially malignant. Although accurate information is not as yet available all available surveys suggest that speckled leukoplakias have a significantly higher rate of malignant transformation than homogenous lesions.

## Treatment of leukoplakias

It is not within the remit of this book to deal in any detail with the surgical treatment of leukoplakias. It should be pointed out, however, that one of the problems in therapy lies in accurately defining the area at risk. It is evident that the oral mucosa as a whole must, to some extent, be exposed to the particular aetiological agent whether the latter be intrinsic or extrinsic. Thus, it is often difficult to come to an arbitrary decision about the limits of surgical excision necessary for optimum treatment. It is by no means uncommon for an apparently generous excision to prove inadequate or for entirely new fields of abnormality to appear. Indeed, it is clear that leukoplakia often represents more than a localized abnormality and that the whole of the mucosa may be involved, even though localized lesions may be the expression of this. Similarly, the recurrence of lesions repaired by mucosal grafts or allowed to epithelialize can be easily explained by the realization that such a manoeuvre is effective only in removing specific epithelial cells which have become involved in a complex abnormality. If the pre-existing aetiological factors remain then recurrent abnormality may be expected even though the aetiology may remain unrecognized. It may well be, in recurrent cases of this nature, that abnormal mesodermal–epithelial reactions are responsible for the initial lesion and are maintained by the presence of the influencing corium after surgery. The same considerations evidently apply to cryosurgical treatment of these lesions and, presumably, account for the few cases of recurrence with malignant change which have been reported. This transformation will depend less on any stimulating effect of the cryosurgical technique than on the maintenance of an abnormal epithelial control mechanism. None the less, there is increasing doubt regarding the recurrence of cryosurgically treated leukoplakias as carcinoma, even though it seems that some degree of under treatment might be involved in such cases. Currently, laser excision of the lesions is felt to be a safer alternative. Certainly, this technique greatly reduces the necessity for grafting of excised areas, there is little tissue distortion following reepithelization and, in terms of post-operative discomfort, it is greatly preferable to cryotherapy.

Apart from surgical treatment it is possible to obtain a marked regression of a significant number of tobacco-induced leukoplakias by discontinuing the habit. Biopsy studies of such regression lesions have shown marked reductions in the incidence of atypia as well as in the extent of the lesions. This manoeuvre should be the first employed in the management of lesions in which a tobacco-mediated aetiology is suspected. Similarly, in those lesions with associated candidal infections, it is often possible to effect a marked improvement by the local use of antifungal antibiotics. It is unlikely that a complete regression will be obtained in this way, but the reduction in the size of the lesion may make the eventual surgical management much simpler. Currently, with the development of safer systemic antifungal agents, trials of their use in candidal leukoplakia are being carried out. The results are not known at the time of writing. There have been several reported attempts to treat leukoplakias with vitamin A, a method based on a quite rational approach since it is known that the vitamin is intimately concerned with the epithelial maturation process and keratinization. It has been demonstrated quite conclusively that large doses of the vitamin can cause regression of fairly large lesions which, histologically, are suspect. However, the improvement has proved to be only temporary, the leukoplakias eventually returning to their original size or even becoming larger on withdrawal of the vitamin. More seriously still, there have been several reported cases of extreme hypertension induced by this treatment and, as a consequence, deaths by cerebrovascular accident have occurred. It must be emphasized, therefore, that this method of treatment, at first sight promising, is one in which the probable dangers far outweigh the probable advantageous effects.

## Pipe smoker's palate

The role of tobacco in the aetiology of leukoplakias in general has been discussed above. Pipe smoker's palate (or leukokeratosis nicotina palati) is quite a different lesion, usually found, as the name implies, in pipe smokers, but also, occasionally, in cigarette and cigar smokers. The characteristic histological feature of this condition is the combination of epithelial acanthosis and hyperkeratosis with inflammatory changes in the mucous glands of the palate. As a result of these changes the palate becomes white and a number of nodules project from the surface, each representing the site of a mucous gland and individually bearing a small red spot as the centre which marks the opening of the duct of the gland (Fig. 7.6). In advanced cases the mucous glands may break down to form quite large chronic ulcers. The condition usually appears most marked on the hard palate although the soft palate may also be involved. It has been suggested that the site of involvement depends upon the projection of pipe smoke directly

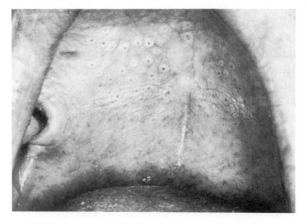

Fig. 7.6. Pipe smoker's keratosis of the palate.

on to the palate, but this is not easy to prove. Oddly enough, in spite of the constant and chronic exposure to the irritant factor in these cases, malignant transformation is very rare and, in general, it is not justifiable to treat this lesion as a premalignant one. Treatment is, in fact, restricted to one method only—to persuade the patient to stop smoking. If this is done, resolution of the lesion (often complete) can be reasonably expected. A covering plate occasionally is a partially successful compromise, but this cannot be of help if the soft palate is involved. There is also the distinct possibility that the plate itself may cause irritation with the superimposition of a condition resembling denture sore mouth on the original lesion.

It should be pointed out that the above description of the condition applies only to that characteristically seen in European and North American pipe smokers. Other tobacco-induced palatal lesions have been described which have a very different prognosis.

## Submucous fibrosis

This is a disease which has only relatively recently been recognized (in 1953), despite presumptive evidence that it must have been present long before that time. The vast majority of cases have been found in the Indian subcontinent, although similar examples have been reported in other Asiatic countries as well as an increasing number of patients in the United Kingdom.

Submucous fibrosis is a condition in which fibrous tissue is laid down in the corium of the oral mucosa. Simultaneous changes occur in the oral epithelium. In the early phase vesicles and small ulcers may be formed, but this stage is soon superseded by one of generalized epithelial atrophy. The effect of the fibrosis is a stiffening of the oral mucosa leading to difficulty in opening the mouth and to a binding down of the tongue. The appearance of the mucosa is of a blanched, marbled nature which seems to be quite characteristic of the condition. It is possible to palpate bands of fibrous tissue within the mucosa and it is reported that eventually the scar formation within the soft palate is sufficient to cause the near disappearance of the uvula. Epithelial atypia are found in a high proportion of the cases and the frequency of leukoplakia is of the order of eight times higher than that in control populations. All studies to date indicate that this condition must be regarded as premalignant although the available evidence is circumstantial. It is true, however, that signs of pre-existing submucous fibrosis are found in a large proportion of oral cancer patients in India, and also that a significant number of patients show epithelial atypia in the oral mucosa so severe as to be histologically classified as 'carcinoma in situ'.

The cause of the condition is not known with certainty, but recent work indicates that both a genetic susceptibility and a fibroblastic response to betel nut chewing may be involved. It is suggested that an autoimmune mechanism is involved: autoantibodies of various kinds have been demonstrated, IgG levels have been shown to be significantly elevated, and the HLA typing shows a preponderance of antigens associated with autoimmune disease. Treatment is difficult and unsatisfactory although many methods have been tried. Intralesional injections of steroids may be minimally helpful in less well developed cases, but surgery (with splitting of the temporalis muscle) may be needed to relieve advanced fibrosis.

## Developmental white lesions

There exist a number of genetically determined white lesions of the oral mucosa, several of which are associated with lesions of other mucous membranes or of the skin. The best known lesion affecting the oral mucosa alone is that of leukokeratosis (white folded gingivostomatitis). The designation 'white sponge naevus' is also commonly used with reference to this condition, but the first use of this term was to describe the oral lesion in a multiple abnormality involving other mucous membranes. However, these terms are often used interchangeably, and it may well be that conditions affecting the oral mucosa alone represent an incomplete expression of a more generalized mucosal abnormality.

Leukokeratosis may be thought of as a naevus, a localized developmental tissue abnormality which is neither inflammatory nor neoplastic. The term 'naevus' is derived from the common pigmented naevus of the skin in which the particular cells involved are the naevus cells (the melanin-producing cells of the epidermis). In some lesions, however, other constituent cells of the skin or mucous membrane may be involved in the abnormality, but the term naevus is also applied to these.

Leukokeratosis appears as a folded white lesion of the oral mucosa which may be so extensive as to affect the whole of the oral cavity. It may be present from birth, but in certain cases the condition may become evident only in later life. There has been, in the past, some doubt as to the exact genetic mechanism involved in the production of this lesion, but it has now been demonstrated that the condition displays classical dominant transmission characteristics in some, but not all patients. In other cases there is no family history and the condition appears to arise in isolation. There are no reported instances of malignant transformation in these lesions, and they are generally regarded as being entirely benign. Histological examination of the affected oral mucosa shows the consistent presence of acanthosis with parakeratosis and intracellular oedema. Characteristic of this lesion and also of many similar lesions in the more complex mucosal syndromes is the presence of many large clear cells extending throughout the stratum spinosum and to the surface of the epithelium (Fig. 7.7). These cells are readily detached by scraping, and a smear preparation of these can be used to demonstrate their characteristic features. Very similar cells are found in smears from other developmental white lesions of the oral mucosa, but minor variations have been described which are considered characteristic of the specific conditions.

A number of syndromes have been recognized in which white lesions of the

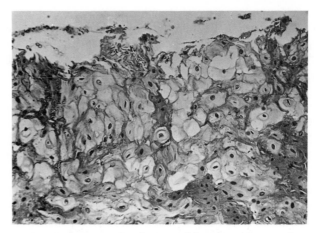

**Fig. 7.7.** Buccal epithelium in a developmental keratosis, showing large clear cells.

oral mucosa are associated with analogous lesions on other mucous membranes and, sometimes with skin lesions. In pachyonchia congenita, for example, white lesions of the oral mucosa and rectum are associated with defective formation of the nails of the hands and feet. In hereditary benign intra-epithelial dyskeratosis the abnormality is restricted to the oral mucous membrane and conjunctiva, but the histology of the oral lesions reveals a degree of dyskeratosis which would be considered alarming in other lesions of the oral mucosa. In spite of this, however, the condition, as its name implies, remains essentially benign. A recently described association is between oral leukoplakia and pre-leukoplakia, oesophageal carcinoma, and tylosis (hyperkeratosis of the palms and soles) (Fig. 7.8). In this example of a complex dermal–mucosal syndrome, the characteristic oral lesion in the young patient is pre-leukoplakia, histologically resembling the lesions of leukokeratosis, but giving way in older patients to lesions better described as leukoplakias which show a less characteristic histological appearance. This latter syndrome, like most other dermal–mucosal syndromes of this type, is transmitted as a dominant condition. It would seem likely that a wide spectrum of such conditions exists and, indeed, isolated examples of varying degrees of combined dermal and mucosal abnormalities have frequently been reported. The named syndromes presumably represent only the most commonly occurring conditions.

## Leukoedema

Leukoedema is the name given to a filmy grey coating of the buccal mucosa which is found in a high proportion of patients. There have been widely divergent estimates of both the incidence and the significance of this condition,

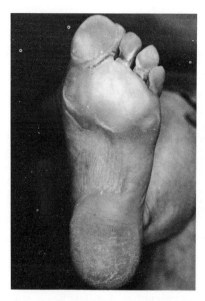

Fig. 7.8. Hyperkeratosis of the feet (tylosis) showing the thickened plantar keratin layer.

one of the difficulties in this being that of definition. However, there can be no doubt that such a filmy coating appears on the buccal mucosa of a large number of completely asymptomatic patients when viewed in adequate lighting conditions. It is probably true that the variation in lighting involved is one reason behind the differing figures obtained by different observers when assessing the incidence of this condition. Similarly, it seems at least a possibility that the higher incidence of this condition reported in coloured populations may be partly due to the increased contrast of colour presented by pigmented mucosae. In some reports it would appear that the lesions described as leukoedema are those which in other surveys would be described as preleukoplakia.

Histologically, this lesion has been described as being parakeratotic, with large swollen cells in the superficial layers of the epithelium. This histological appearance is quite compatible with the observations that the grey surface film (consisting of the superficial oedematous cells) may easily be scraped away from the mucosa, leaving an apparently intact surface which again rapidly acquires the superficially grey appearance.

In the same way that the reported incidence of leukoedema in differing 'normal' populations is very variable (from about 1 per cent to over 50 per cent) the opinions offered as to the effect of external irritants on the condition are divergent. In some reports a relationship between tobacco habits and the incidence of leukoedema has been described, whilst in other reports it has been concluded that leukoedema represents only a variant of normal mucosa, the filmy appearance being due to an incomplete desquamation of superficial parakeratinized cells. All authors are agreed, however, that leukoedema is not associated with epithelial atypia and that it should not be regarded as a premalignant lesion.

# Inflammatory overgrowths and naevae

All new growths of the oral tissues should be treated with suspicion and must be fully investigated. There are some well-defined conditions (for instance, denture granulomas) of which the clinical diagnosis can be accurate in practically all cases, but appearances can be misleading and it cannot be too strongly stressed that absolute certainty as to the nature of any lesion can only come after histological examination.

## Hamartomas

A hamartoma is a localized non-progressive tissue abnormality resulting from a defect in development. It is neither inflammatory nor neoplastic in nature, but since it may be confused with either, it should be considered in relation to them. When present on skin or mucous membrane the term 'naevus' is often used even when the naevus cells (the melanocytes) are in no way involved. In such a lesion, a single element of the mucosa, epithelial, vascular, or lymphatic, is predominantly involved. An example is the very common pigmented naevus (mole) of the skin which involves abnormal development of the melanin-producing cells in the area. Leukokeratosis has been discussed above; this may be considered as an epithelial naevus but it should be added that this concept of the lesion, has recently been challenged by Pindborg et al., who consider the lesion to be little different from other leukoplakias.

## Angiomatous naevae

These are the result of developmental abnormalities in either the lymphatic or vascular components of the mucosa. Vascular naevae (haemangiomas) are relatively common lesions of the oral mucosa resembling the 'strawberry mark' of skin. Depending on the degree of dilatation of the abnormal blood vessels, the lesion may appear as a fine network of capillaries or as a more pronounced nodular structure, usually filled with slow-moving venous blood and, therefore, dark blue in colour. The cavernous form may be mistaken for a melanoma, but can be quickly differentiated by its tendency to blanch on pressure. This can be seen by pressing a glass slide down onto the surface. It should be emphasized that the vascular naevus is a static developmental abnormality which is asymptomatic and is best left undisturbed. A similar naevus involving the lymph vessels is the lymphangioma. It consists of a collection of dilated lymphatic vessels and spaces in a connective tissue stroma and is very similar to the vascular naevus in structure. Although it can appear on any part of the oral mucosa it is seen much more commonly on the tongue than elsewhere. If the lesion is superficial it appears as a translucent white structure on the mucosa, if it is deeply situated the overlying surface of the tongue appears greyish and nodular. This, too, is an entirely innocuous lesion. Occasionally, the tongue is the site of mixed lesions: lymphhaemangiomas in which there are elements of both types of lesion in close relationship.

There may be problems of bleeding as a result of trauma to these lesions and action may be called for, although the bleeding rarely reaches dangerous proportions in these circumstances. Cryosurgery has been used to deal with this situation in the recent past, although not without some difficulties; there can be problems with post-operative oedema of the tongue. Presumably, the rapid development of lasers will provide an answer to this problem. Any form of surgical interference with these lesions should be carried out in the hospital environment.

If there is any doubt whether or not a lesion is a naevus, biopsy examination may be necessary, but angiomatous lesions often bleed copiously. No obviously vascular lesion should ever be subjected to any form of surgery except in hospital.

## Epulides

An epulis is defined as a soft tissue swelling of the gingival margin. The term is more specifically used to describe a range of hyperplastic inflammatory lesions arising from the periodontal tissues. They represent an exaggerated inflammatory response of the periodontum, although the source of irritation is not always obvious. The inflammatory process arises initially in the interdental tissues and there is often an associated loss of alveolar crest-bone which may become severe in advanced cases. Two types, the fibrous epulis and the giant-cell epulis, are commonly seen, reflecting differing stages of the inflammatory and bone resorbing process.

A fibrous epulis is, essentially, heavily fibrosed granulation tissue. Its content of collagen fibres gives it a firm, rubbery texture and its colour is pale pink (Fig. 7.9). However, the degree of fibrosis depends on the stage of maturity of the epulis and a lesion in its early stages may be soft in texture and with a histological appearance that shows very many cells; in a long standing lesion calcification may occur. Acute inflammatory changes may follow trauma or infection and in such cases, the epulis may become red and sore. There is a preponderance of females among patients presenting with fibrous epulis, the female/male ratio being variously quoted in different surveys as from 4:1 to less than 2:1. This discrepancy in the statistical material reported extends to other factors; it is difficult to see why different centres report such variable results. The only consistent factor is the predominance of the vascular epulis (discussed below) in female patients of child-bearing age.

A variant of the fibrous epulis is the vascular epulis or pyogenic granuloma.

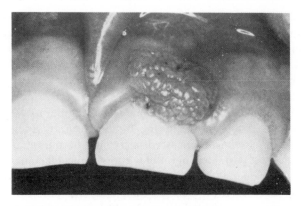

Fig. 7.9. Fibrous epulis. The nodular appearance is a little unusual.

This latter misnomer is still widely used and refers to an old concept that these lesions, and similar ones of the skin, are a response to infection by pyogenic bacteria. It is essentially the same as the pregnancy epulis (Chapter 9).

In this lesion, the granulation tissue remains vascular and immature. It is much redder in colour than the fibrous epulis proper (Fig. 7.10) and has a tendency to bleed easily because of its high vascular content. However, it is essentially the same lesion as the fibrous epulis, differing only in the maturity of the tissue of which it is composed.

The giant cell epulis is a lesion in which the granulation tissue is osteogenic in nature. Its predominant histological feature is the presence of multinucleated giant cells dispersed in a vascular stroma. With maturity the lesion may become less vascular, more fibrosed and may include some areas of bone formation. In its immature form, this epulis is characteristically red–purple in colour. The reported distribution in terms of age and sex varies even more widely in the case

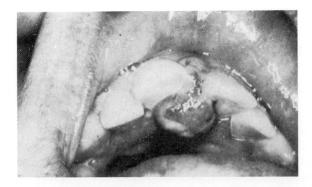

Fig. 7.10. 'Pyogenic granuloma', an epulis consisting of immature granulation tissue.

of the giant cell epulis than in the fibrous epulis. Perhaps these discrepancies are the result of the well known difficulties in assembling representative samples on a retrospective basis.

Treatment of all these forms of epulis is by local excision. The origin of the lesion is often interdental and, in more advanced cases, the periodontal membrane may be quite deeply involved. If excision is not complete, there may be recurrence of the lesion and so, although radical surgical techniques are not called for, the initial removal should include all affected tissue. With repeated recurrence, it is sometimes necessary to remove the adjacent teeth in order to secure the elimination of the tissue of origin. In their tendency to recur, epulides may appear to be neoplasms, but the recurrence is due only to persistence of the conditions which caused the initial abnormal response.

Although the clinical diagnosis may be a confident one it should invariably be confirmed by histological examination; occasionally, a neoplasm may present in a form resembling a simple epulis and in a likely site for one. It should be remembered that there is an occasional relationship between the occurrence of a giant-cell epulis and the presence of hyperparathyroidism. It would probably be wise to carry out appropriate investigations (including blood tests for plasma calcium, phosphorus, and alkaline phosphatase) on all patients presenting with a giant cell epulis. Any symptoms or history which might imply undiagnosed hyperparathyroidism (such as renal calculi) should be taken into consideration.

## Fibro-epithelial polyp

This lesion, similar in structure to the mature fibrous epulis, is essentially scar tissue produced as a response to trauma such as may result from repeated irritation of the buccal or labial mucosa, along the occlusal plane of the teeth, often casued by a bite. It is usually seen in adults and there is no sex differentiation. The lesion appears as either a sessile or a pedunculated swelling (Fig. 7.11) and is quite free of symptoms unless secondarily traumatized. The

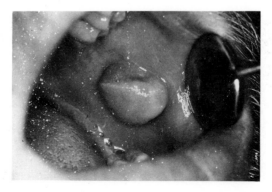

Fig. 7.11. A large fibro-epithelial polyp of the buccal mucosa.

usual size of such a lesion when the patient presents for treatment is of the order of 1 cm in diameter, but occasional long standing lesions are seen which are very much larger. The colour of the lesion is pink and the texture varies from soft to rubbery, depending on the maturity of the constituent fibrous tissue. Since this lesion is simply an exaggerated and chronically irritated mass of scar tissue, treatment need only be conservative, excision to the limit of the swelling, or to the base of the pedicle being all that is required. Recurrence will occur only if trauma is repeated and is, in fact, uncommon. As with all other tissue overgrowths, absolute certainty of diagnosis can come only after histological examination although a clinical diagnosis can often be made with a fair degree of confidence.

## Denture granuloma

This is essentially a similar lesion to the fibro-epithelial polyp, the irritating factor in this case being the flange of an over-extended or ill-fitting denture. As in the case of the polyp, proliferative scar tissue is formed following chronic trauma, the typical fissured shape of the denture granuloma depending on the indentation caused by the flange of the denture (Fig. 7.12). These lesions are rarely painful and, indeed, often cause astonishingly little trouble to the patient. This being so, occasional lesions are seen which are very large indeed, with multiple folds of proliferated tissue. It is, as yet, undecided whether these lesions ever become malignant, but it can be said with some confidence that malignant change, if it does occur, must be very infrequent indeed. It is certainly justifiable to consider the denture granuloma as an entirely benign lesion and to treat it by simple excision after removing the offending denture or drastically trimming it away from the affected area. The removal of the source of chronic irritation is in many cases sufficient to reduce very considerably the size of the lesion within a relatively short time and even to make excision unnecessary.

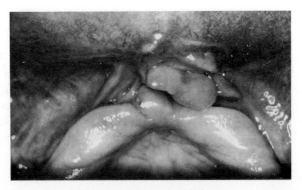

Fig. 7.12. Denture granuloma.

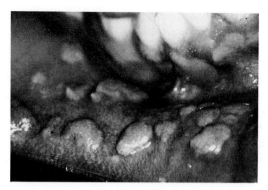

Fig. 7.13. Labial lesions in focal epithelial hyperplasia.

## Focal epithelial hyperplasia

This condition (otherwise known as Heck's disease) has only recently been fully described and investigated. The patients are predominantly children from coloured African, Eskimo, and American Indian groups, although a very few cases have been reported in white Europeans. In this condition multiple raised sessile lesions appear on the buccal and labial mucosa (Fig. 7.13). The mucosa retains a relatively normal pink appearance and the texture of the lesions is soft. There is no ulceration (except in the case of secondary trauma) and the lesions are quite pain free. Histologically, the epithelium, overlying a relatively normal corium, appears hyperplastic, with marked cellular irregularities. It is suggested that these changes are consistent with a viral aetiology for the condition, although this is not yet proved.

No treatment is necessary; it is a self-limiting condition and regresses completely after a variable period. The possibility of diagnostic confusion (on a clinical basis) with oral Crohn's disease (Chapter 8) should be borne in mind.

## Neoplasia

Neoplasms of many kinds, both primary and secondary, may occur in the mouth. It is possible to describe typical clinical features for many of these lesions, but it must be remembered that misleading and atypical forms may also present and may lead to diagnostic error. The final diagnosis of any tissue overgrowth depends entirely on histological investigation and this must invariably be undertaken if any doubt at all exists as to the true nature of a lesion.

### Papilloma

This benign epithelial neoplasm is relatively common. It may appear anywhere on the oral mucosa, but is most commonly found at the junction of the hard and soft palates. Its typical appearance is 'cauliflower-like' and pedunculated (Fig. 7.14) with a pale colour varying from that of normal mucosa to white; it is a painless lesion which rarely gives trouble. There is little evidence of malignant transformation in these lesions and, in this respect, the oral lesions behave quite differently from those in the lower parts of the gastro-intestinal tract which have a definite tendency to undergo malignant change. An interesting clinical variant is one in which the papilloma may develop under the palate of an upper denture. In such cases the normal shape of the papilloma is distorted and flattened into a thin disc which fits into a shallow depression in the palate. The papilloma retains its thin pedicle and may be displaced downwards on it like a hinged flap.

Treatment of a papilloma is by local excision. This, however, must be sufficiently wide and deep to include any abnormal cells which may extend beyond the area of the pedicle. It is not sufficient to simply sever the pedicle; this will lead to recurrence, cryotherapy is an alternative.

Multiple viral warts, clinically similar to the solitary papilloma, occasionally occur in the mouth or, more commonly, on and around the lips. In children these are often the result of autoinoculation by chewing warts on the hands. Sexual transmission between genital and oral sites has also been described. A large number of papillomaviruses has been described and it is possible to classify oral warts/papillomas in various ways. For practical purposes, however, the division is clear between solitary lesions with no evident infective aetiology and those the result of inoculation. Management is the same, although clearly the wart-chewing habit of affected children should be discouraged. Viral warts generally spontaneously regress after a period of a year or so.

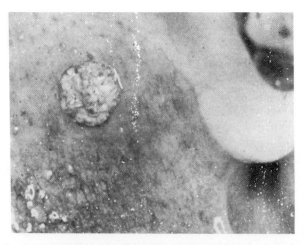

Fig. 7.14. Papilloma of the palate.

## Salivary gland tumours

Occasionally, neoplastic growth occurs in the minor salivary glands of the oral mucosa. Since the greatest concentration of these glands is in the area of the junction of the hard and soft palates, this is the region in which these neoplasms are most often seen. Their growth is usually slow and painless and ulceration is unusual unless there has been some degree of trauma (Fig. 7.15). The texture of such a tumour is usually firm and the overlying mucosa may appear virtually normal.

The outstanding characteristic of these growths is their unpredictability, both in histological appearance and in clinical behaviour. Most of them may be considered, at worst, as locally malignant and, indeed, many act in a virtually benign manner. The predominant histological findings is of pleomorphic adenoma, but a wide range of structures may be demonstrated, including that of adenocarcinoma and other aggressive lesions. The clinical behaviour is correspondingly unpredictable, some of the more aggressive lesions being entirely malignant, leading to death both by local tissue invasion and the production of distant metastases. The only clue as to the probable eventual behaviour of a salivary neoplasm is from the histological examination of tissue and this must form the basis of the surgical treatment of the lesion.

It is clear that the treatment of such a salivary tumour is outside the scope of general practice, but equally clearly, the initial recognition of the tumour must often be the responsibility of the dentist. These lesions, because of their slow and painless growth are often very deceptive, and lack of complaint from the patient may produce a quite unwarranted sense of security. Any suspected salivary tumour should be referred at once for investigation and treatment without any attempt at minor surgical intervention which might make later definitive treatment difficult.

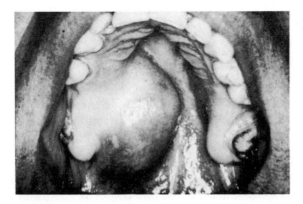

Fig. 7.15. Salivary gland tumour in the palate (a pleomorphic adenoma).

## Carcinoma

There is great variation in the incidence of oral cancer in differing parts of the world. In India, for example, 40 per cent of all cancers occur in the mouth, whereas in England the figures are very much less, of the order of 2 per cent. The prognosis for the patient with oral cancer is in general poor and much depends on early diagnosis. In this respect the role of the dental practitioner in recognition of the early lesion is vital.

Carcinomas may occur in any area of the oral mucosa, although the sites particularly at risk depend on a number of factors. Of these, the most important is the use of tobacco in any of its forms. As an example of this, the region of India in which reverse cheroot smoking is practised amongst women shows a vastly increased incidence of carcinoma of the palate in female patients. In Europe, such irritating factors are usually less obvious, but there is no doubt that tobacco smoking and chewing according to the European pattern can be the source of epithelial dystrophy. However, little is known of the relationship between intrinsic and extrinsic factors in the production of malignant disease. As has been pointed out, there are undoubtedly patients in whom intrinsic epithelial instability is present and, in fact, a substantial number of oral carcinomas appear to be multicentric in origin. One of the greatest surgical problems in the management of oral malignancy arises where a patient has evident gross abnormality of practically the whole of the oral mucosa.

Not all oral carcinomas are preceded by a recognizable premalignant lesion, although in many cases evident abnormality may have been present for some time. The significance of leukoplakia as a premalignant lesion has been discussed above but it must be repeated that the clinical criteria by which a white patch may be judged as premalignant are far from clear and that it is only after a study of the histological appearance of the lesion that any attempt at prognosis can be made.

The classical signs of a malignant condition, which should arouse immediate suspicion are:

(1) persistent ulceration: any unexplained ulcer which lasts for longer than 10 days should be treated with suspicion;

(2) induration: thickening and hardening of the tissues;

(3) proliferative growth of tissue above its normal level, often with changes in the surface texture and colour changes;

(4) fixation of the affected tissue to the underlying structures.

Pain is not always present in the early stages of a carcinoma and the patient may be quite unaware of any abnormality until the lesion has become large and secondarily infected. If the affected area includes the teeth these may become mobile due to the replacement of the periodontal membrane by tumour. Thus, unexplained rapid loosening of the teeth should be most carefully investigated.

Lymph node involvement may occur early in oral carcinoma. However, enlarged regional lymph nodes may in fact show only non-specific inflammatory changes on biopsy in some early cases. Unfortunately, this cannot be relied on and, in some cases, malignant deposits in the lymph nodes are found whilst the primary lesion is still small. It is certainly true that prognosis is more favourable if treatment is instituted before the lymph nodes are in any way affected.

Lesions showing the signs outlined above are immediately suspect but such a lesion is at a relatively late stage (Fig. 7.16). In its early stages, a carcinoma may show none of these signs and may be detectable only by a change in colour or surface texture of the mucosa. It may be impossible to differentiate clinically between leukoplakia or erythroplakia and early carcinoma and, thus, biopsy examination is mandatory in every case in which the possibility of carcinoma may arise (Fig. 7.17). Any prolonged ulcer, erythema, or unexplained white patch should be investigated in this way. As has been explained in Chapter 2, smear biopsy is entirely unsatisfactory for the diagnosis of suspected oral malignancy. Incisional or excisional biopsy followed by tissue examination is essential for a proper diagnosis to be made. A diagnosis of carcinoma should be followed as rapidly as possible by treatment and so it is essential that any dental surgeon who carries out biopsy procedures should be in a position to arrange for immediate treatment if required. For this reason, it is often better for such investigation to be carried out in specialist centres rather than in the practitioner's surgery. Prognosis in late oral carcinoma is, as has been pointed out, poor. Much depends on the early detection of the lesion before the cervical nodes become involved, the overall five-year survival rate being effectively doubled if the lesion is discovered early. It is also true that the prognosis becomes worse the further back in the oral cavity the lesion lies. Thus, it is quite clear that the patient's survival may depend on a careful and critical soft tissue examination by the dental surgeon.

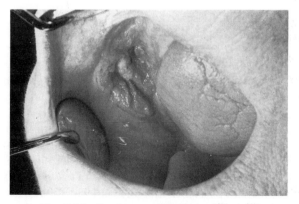

Fig. 7.16. Carcinoma of the buccal mucosa.

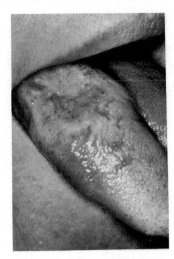

**Fig. 7.17.** Early carcinoma of tongue

In order to aid uniformity in studies of the epidemiology of oral cancer and the prognosis of those suffering from it there have been a number of assessment methods introduced. These are generally known as staging systems. The most widely used is the STNMP system (Site, Tumour, Nodes, Metastasis, Pathology). In this system gradings are given to each patient dependant on such factors as the size of the primary tumour, the extent of regional lymph node involvement and the presence or otherwise of distant metastases. These systems are for use in the study of patients with advanced neoplasms and to aid the planning of the major treatment regimes which might be required. They have been found of considerable value in assessing the overall ultimate prognosis of patients in each group although, clearly, the stage group to which a patient is assigned represents a range of possible subsequent clinical progess.

Cancer of the lip is a rather special lesion in that it is clearly visible and so is noticed early. It has a much better prognosis than intra-oral cancer and early diagnosis often implies successful and relatively small-scale surgery. The lower lip is practically always that affected, the patients being predominantly older males (Fig. 7.18). The carcinoma is often mistaken in its early stages for a herpetic lesion, but its persistent nature should arouse suspicion and the same criteria for investigation should be adopted as for intra-oral lesions.

A wide variety of malignant lesions other than carcinomas may occur in the oral cavity but are very much less common than is carcinoma. Early diagnosis depends on the maintenance of a high degree of awareness by the dental surgeon together with an insistence that any clinically undiagnosable condition should be subjected to complete investigation.

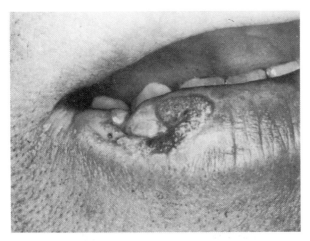

**Fig. 7.18.** Carcinoma of the lower lip.

## Radiation mucositis

The treatment of oral carcinoma is a complex matter which may involve surgery, chemotherapy, or radiation. If radiation is used it is almost certain that the oral mucosa as a whole and the salivary glands also will be affected.

During the course of radiation (which is likely to last for several weeks) the patient develops an increasing generalized erythematous and ulcerative response of the oral mucosa (radiation mucositis). This is extremely painful, the problem being complicated by the fact that the salivary glands are also virtually always affected, with consequent lessening of salivary secretion. Secondary candidiasis almost invariably occurs. Following the course of radiation the mucositis gradually subsides leaving an atrophic oral epithelium on a relatively avascular submucosa. The salivary flow may be permanently reduced.

In the acute phase, simple bland mouthwashes may help in the maintenance of oral hygiene. Lignocaine mouthwashes are helpful before meals and miconazole oral gell may be used to combat candidiasis. In the long term the combination of dry mouth and an atrophic mucosa may cause continuing discomfort for which artificial salivas are the only available treatment. Those containing added fluoride might well be considered; these patients become unusually susceptible to caries against which topical fluorides seem particularly effective.

# 8

# Diseases of the skin: gastro-intestinal disease

## Diseases of the skin

It has been pointed out in Chapter 1 that the oral mucous membrane, although similar to other lining mucosae in structure and general behaviour, also resembles the skin in some ways. This is in keeping with the transitional position and function of the oral mucous membrane, lying as it does in an intermediate position between the skin and the gut mucosa proper. One result of this situation is that generalized diseases, both of mucosae and of skin, may affect the mouth. However, in such skin diseases the oral lesions, because of the modifying effect of the environment, may bear little superficial resemblance to those of the skin. Not only the continual presence of saliva, but also the certainty of secondary infection by oral organisms and the repeated traumas of the oral environment play their respective part in the modification of the oral lesions. Particularly good examples of this form of modification are shown in the group of diseases in which blisters or bullae are formed. When this occurs in the mouth the bullae rapidly break down to form ulcerated areas. Such behaviour affects the management of these conditions; it is often necessary to treat the secondary infection produced in this way before proceeding to more systemic treatment. A different form of modification in oral lesions occurs in some diseases (for example, lichen planus) in which the initial oral lesion may bear little morphological resemblance to the skin lesion, even though the basic histopathological changes are similar. These variants presumably depend on the differing structures of the skin and the oral mucosa rather than on the effect of the environment.

It is not easy to understand why some skin diseases commonly produce oral lesions whilst others do not. For example, psoriasis, one of the two most common skin diseases seen in European clinics (the other is eczema) produces oral lesions very rarely; in fact, there is some difference of opinion as to whether a characteristic oral lesion does exist. On the other hand, lichen planus produces oral lesions in a high proportion of the patients involved. Oral lesions have occasionally been described in a very wide range of skin diseases, even though such behaviour may not be characteristic of the condition. Thus, in the diagnosis of oral lesions of doubtful aetiology, a history of skin diseases is always of interest. Perhaps the most important consideration in this context is the fact that in some diseases of the skin (including some of the most serious such as pemphigus) the oral lesions may appear before those of the skin and may thus provide an opportunity for early diagnosis and the rapid initiation of treatment.

Definitive diagnosis of the oral lesions in skin diseases depends largely on biopsy examination and this should be carried out on any doubtful lesion. However, this is not always an easy matter, particularly in the case of bullous lesions in which the manipulations of incisional or excisional biopsy taking may cause virtual disintegration of the lesion. If such a lesion cannot be removed intact for examination then the edge of the lesion may often show the characteristic changes better than a central area where secondary changes may obscure the picture, a biopsy taken from an ulcerated area may show nothing but non-specific inflammatory changes. In a very few instances, notably in the bullous diseases, cytological diagnosis may be carried out by the examination of loose cells aspirated with bulla fluid or scraped from the edge of a burst lesion. However, immunofluorescent techniques and, in particular, direct immuno-fluorescent studies of biopsy material have revolutionized the methods of diagnosis of skin diseases and their oral lesions. The selection of biopsy sites and the method of handling of the specimens are quite different than in the case of biopsies taken for routine histology (these factors are discussed in Chapter 2).

In this chapter the conditions chosen for discussion are those which are relatively commonly seen in the oral medicine clinic or those of particular interest or importance in the dental context.

## Lichen planus

In this disease the skin lesions take the form of dusky pink papules which may occur in any site, but are most commonly found on the front surfaces of the wrists, on the genital skin, on the abdomen and lumbar regions, and on the neck. Fine white striations overlie the papules, these are so-called Wickham's striae which are characteristic of the condition (Fig. 8.1). In general, the skin lesions are relatively short lived, the average duration being of the order of 9 months; after this they fade, leaving behind a faintly pigmented patch which may take a considerable time to disappear. However, in a significant proportion of the patients there may be a recurrence of the lesions. The majority of skin lesions cause little trouble apart from itching which is very variable in its intensity, in some instances it is so insignificant that the lesions are not noticed by the patient. In a few cases the Koebner phenomenon may be seen with lesions distributed in a linear pattern along a scratch mark on the skin. Scalp lesions may also occur in a few patients (usually female); often these are not papular, but are represented by patches of alopecia. There are widespread variations in the clinical picture shown by occasional patients who may present with bullous or hypertrophic lesions and, very rarely, the disease may first occur in an acute form with initial symptoms much more severe than those described above. In a

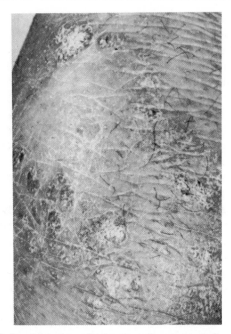

**Fig. 8.1.** Lesions of lichen planus on the forearm showing papules and Wickham's striae.

minority of patients (but still a substantial number) the lichen planus is precipitated by one of an ever widening range of drugs. This is termed a lichenoid reaction, but there seems little difference between the drug-induced and non-drug-induced condition. In particular, the lesions do not, by any means, uniformly regress on withdrawal of the drug. Non-steroidal anti-inflammatory agents, antihypertensive drugs, lithium, gold injections, antimalarials, and various antibiotics are among the drugs which have been implicated. An industrial chemical which may have the same effect is included in the developer used for colour films. When induced in this way the lesions are identical in all respects (including their histology) with those of the non-drug-induced forms. Most patients with skin lesions are between the ages of 30 and 60 years when first presenting for diagnosis and rather more than half are female (54 per cent in a recent survey). It has for long been suggested that the patients are drawn predominantly from those with managerial or professional responsibilities, but investigations have not substantiated this.

The histology of a skin lesion is quite characteristic and, in all essentials, resembles that of the oral lesions. The most obvious feature is the presence of a narrow dense band of inflammatory cells, predominantly lymphocytes, within the dermis and lying just below the epithelium. In the skin the effect appears to be a pushing up of the epithelium to form the papules although this may not, in fact, be the actual mechanism of papule formation. The overlying epithelium may undergo a variety of changes, hyperorthokeratosis being the most common finding in the case of the skin lesions.

Oral lesions occur in a considerable proportion of patients with lichen planus and are seen in the oral medicine clinic far more frequently than the oral lesions of other skin diseases. The true incidence of oral lesions is a little difficult to determine since the figures available vary widely, a significant factor being whether the patients are first seen in a dermatological clinic (in which case oral lesions are present in about 70 per cent of cases) or in a dental clinic. In these latter patients, presenting because of oral lesions, the incidence of skin lesions is of the order of 40 per cent. This variation is probably due to a number of factors, the most important being the asymptomatic nature of the oral lesions in many cases. A second important factor is the inconstant sequential relationship of the lesions; the oral lesions may occur before, after or at the same time as the skin lesions. In general, however, oral lesions last much longer than skin lesions, a mean duration of 4.5 years has been suggested but there is no doubt that in many cases this period may be greatly exceeded. In the case of oral lichen planus most patients presenting are female—70 per cent in a number of reported series. The age range is similar to that in patients reporting with skin lesions only although in the case of patients with oral lesions there is a rather higher proportion of patients in the 60 year plus groups. The author has seen a small number of patients with confirmed oral lesions at a very early age, the youngest being 7 years old. This would be considered a great rarity in the case of skin lesions. The oral lesions are usually bilateral and involve the buccal mucosa in practically all cases. In descending order of frequency the tongue, gingivae, palate, and lips may also be affected.

The essential histological change in these lesions is the same as in those of the skin, the presence of a subepithelial band of inflammatory cells (Fig. 8.2), but in the case of the oral mucosa there is a much wider range of epithelial response, with acanthosis or atrophy, orthokeratosis, or parakeratosis. However, the most common finding is of parakeratosis, the rete pegs being distorted to give a 'saw-tooth' appearance or, more commonly, flattened. Around the basement membrane there is oedema, associated with degenerative changes in the basal cells, an association which in some cases leads to virtual separation of the epithelium from the corium (Fig. 8.3). The direct immunofluorescence findings in lichen planus are not highly specific. A band of fibrin is shown up at the basal zone, but no immunoglobulin deposits. Cytoid bodies may be seen, however, both in the epithelium and in the dermis. These are non-disease-specific spherical structures which contain variable immunoglobulins and complement components. A high incidence of these is indicative, if not diagnostic, of lichen planus.

Depending, largely, on the precise nature of these epithelial changes, a wide range of oral lesions may occur. It is possible to describe these lesions in a number of ways, but for clinical purposes a classification into three groups provides a useful guide to behaviour. These groups are: non-erosive lichen planus, minor erosive lichen planus, and major erosive lichen planus.

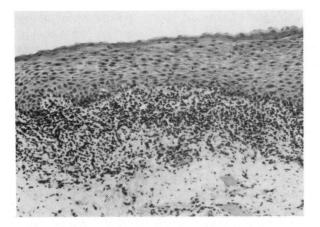

**Fig. 8.2.** Section of oral mucosa in non-erosive lichen planus showing flattening of rete pegs, parakeratosis, and a band of chronic inflammatory cells in the corium.

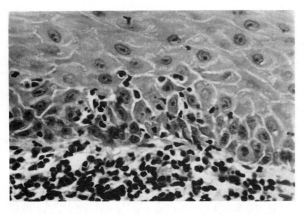

**Fig. 8.3.** High-power view of basal cell layer of oral mucosa in lichen planus showing degenerative changes.

## Non-erosive lichen planus

In this variant the epithelial change is of hyperparakeratosis, or occasionally, hyperorthokeratosis. There is no atrophy of the epithelium and, hence, no ulceration. In recent surveys approximately 30 per cent of cases fall into this group although in some series the proportion was reported as being much higher. For example, Cooke in 1957 reported a 75 per cent incidence of non-erosive forms in a series of cases, but in 1967 found a much lower incidence, a difference explained as possibly being due to a change in the form of the disease.

An alternative possible explanation would be dependent on a change in the intake of patients. It is possible that in earlier series the more markedly symptomatic patients were seen in clinics other than dental clinics, a situation which is no longer likely to apply.

The characteristic appearance is of white streaks on the oral mucosa, arranged in a reticular pattern (Fig. 8.4). In some patients the lesions may be more confluent and may resemble a leukoplakia whilst in others there may be papular, linear, or annular arrangements of the white areas. These are often described as being similar to the Wickham's striae of the skin lesions, but are, in fact, much more clearly defined. In the course of the disease the morphology of the lesions may change, and there may also be variations in their extent and intensity. In general, these lesions are quite symptom free and are often noticed incidentally by the patient although a sensation of roughness may be present. In the striated forms the diagnosis may initially be made with some confidence on appearance alone, but in the case of the confluent leukoplakia-like lesions the diagnosis may be made only after biopsy.

### Minor erosive lichen planus
In this form of the disease the oral epithelium undergoes atrophic changes and is easily lost from the weak and oedematous basal areas. The result of this is the formation of shallow erosions on the mucosa, these often being associated with nearby areas in which non-erosive lesions occur. In a few patients the basal and subepithelial oedema may lead to separation of the epithelium with consequent bulla formation, but such bullae, developed in initially atrophic epithelium, are very fragile and rapidly disintegrate to produce the characteristic erosions of the condition. Thus, this clinical grouping (minor erosive lichen planus) includes those variants which might be described as atrophic, bullous, or ulcerative.

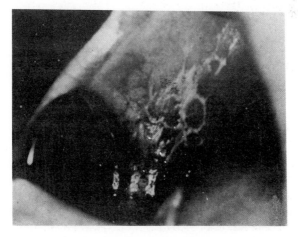

Fig. 8.4. Reticulated non-erosive lichen planus of buccal mucosa.

Unlike the situation in the non-erosive form of the disease there is often considerable discomfort to the patient, particularly when eating spicy or acid foods. The mucosa is also susceptible to mechanical irritation and the initial symptoms may occasionally appear as denture trauma before any other more characteristic lesions appear.

As in the non-erosive form of the disease, the definitive diagnosis is by biopsy. Although the appearance of the lesions may give an indication of the diagnosis in some cases, this is not always so and the clinical features may sometimes be confused with those of pemphigus or other bullous lesions or with patchy erythroplakia. The presence of skin lesions is a useful diagnostic pointer but, evidently, not a constant one. Many of the cases described as desquamative gingivitis are, in fact, minor erosive lichen planus involving the gingivae although pemphigoid may present in this way (Fig. 8.5).

## Major erosive lichen planus

This variant of oral lichen planus represents a clinically distinct form of the disease which is relatively uncommon. It is usually of sudden onset and the dominant feature is ulceration of the tongue, buccal mucosa, and labial mucosa. The ulceration is widespread and, in particular, may involve a large part of the surface of the tongue. This ulceration may occur with little prior warning and is associated, at least initially, with only minor areas of non-erosive lesions; often the only white lesion visible is at the margins of the lips. The ulcers have a glazed appearance and are separated from the adjacent mucosa by a clearly demarcated edge (Fig. 8.6). Following the acute onset of the condition there is usually a period of little change lasting some months and, after this, a period of slow resolution which may be as long as two years or so. With this healing process the ulcers are replaced, not by normal mucosa, but largely by white confluent lesions. During the acute phase the patient may experience severe discomfort

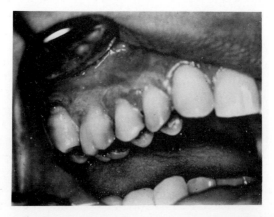

**Fig. 8.5.** Minor erosive lichen planus affecting the gingivae ('desquamative gingivitis').

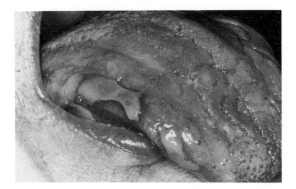

**Fig. 8.6.** Major erosive lichen planus of the tongue.

and find it difficult to eat solid food, but the pain complained of by the patient is rather less than might be expected in view of the widespread nature of the lesions. Most patients with this form of lichen planus are in the older age groups, but no consistent associated systemic disturbances are found on investigation. This is a most intractable condition and there is little to be done to influence its progression. However, it can be stated with some degree of confidence that slow resolution (at least to the stage of non-erosive lesions) is likely to occur spontaneously.

The great majority of cases oral lichen planus run a completely benign course but, in a very small proportion of cases (certainly less than 1 per cent), malignancy may supervene. For this reason, long-term reviews of all cases of lichen planus should be carried out and biopsy (or re-biopsy) performed if any evident clinical change should occur. It has often been suggested that it is in the erosive variants of lichen planus that malignancy is likely to occur and there have, in fact, been a few such cases described with supporting histological evidence. However, there are no controlled studies of this subject.

The aetiology of lichen planus is obscure and there is, as yet, no convincing evidence of the process involved. A viral origin has been suggested by several workers, but the intranuclear structures found in the lesions and initially considered to be of viral aetiology have subsequently been shown to be entirely non-specific and to occur in a variety of epithelial disorders. The immunological changes in the condition have been described by Lehner who finds that IgG levels are suppressed whilst IgA levels rise, IgM remaining unchanged. This is evidence against an autoimmune origin for lichen planus since, in the autoimmune diseases, IgG levels are invariably raised. However, more recent work based on the study of the immunological characteristics of the epithelial cells in the lesion has implied the possibility of a local autoimmune response to the epithelium. It is suggested that the Langerhans cells may play a prominent part in this process. The majority of the lymphocytes in the subepithelial infiltrate have been shown to be of the T type, with cytotoxic T cells in the epithelium, facts consistent with an autoimmune mechanism. Comparative studies of the immunologically competent lymphocytes in the dermal infiltrates in lichen planus and in leukoplakia have shown considerable differences between the two conditions and, although no firm conclusions as to the aetiology may be gained from this work, it is evident that there are major differences in the immunological mechanisms operating in the two situations.

As has been pointed out, the most characteristic histological finding in lichen planus of either skin or mucous membrane is the zone of inflammatory cells within the dermis, a feature in all lesions, whatever the overlying epithelial response. It has been shown that in patients with lichen planus of the skin, but without oral lesions, such changes can be found in the clinically unaffected oral mucosa.

It would seem that the mesodermal changes are widespread even though the epithelial abnormalities may be localized. This would seem to be an argument for the mesodermal origin of lichen planus. All in all the question of the aetiology of the disease cannot yet be said by any means to be settled.

The treatment of generalized lichen planus is by the use of steroid creams, which may be effective in reducing the itching and irritation. It is generally accepted that no form of treatment is likely to greatly influence the duration of the disease. In a very few cases, when irritation is intense, systemic steroids may be given in an attempt to reduce the discomfort, but apart from all other disadvantages associated with the use of such therapy, the production of hypertrophic lesions may be induced. Should this form of treatment be used the steroid must eventually be reduced with great care since a sudden cessation of the treatment can cause a severe rebound exacerbation.

Treatment of oral lichen planus depends largely on two procedures: the use of antiseptic or antibiotic mouthwashes to relieve severe discomfort caused by the presence of ulcers and erosions, and the use of local steroids to assist in the healing of erosions. Of the available steroids betamethasone-17-valerate is the most effective in the majority of cases. Although there is no oral preparation generally available the use of an aerosol spray (designed for the treatment of respiratory disease) delivering a dose of 100 μg per application three or four times daily may be very effective in patients with the minor erosive form of the disease. If this steroid is used a watch must be kept for the onset of oral pharyngeal or laryngeal candidiasis. Any sudden worsening in the soreness of the mouth or the onset of hoarseness in a patient under treatment with this steroid should be suspected as being due to candidiasis and treated by antifungal therapy. Miconazole is particularly effective; the use of the oral gel as a mouthwash-gargle will usually eliminate the laryngeal symptoms as well as those in the mouth. In a few patients with painful lesions of the lips a light application of betamethasone-17-valerate cream may be of value. As has been pointed out, major erosive lichen planus is an intractable condition. Local steroids in the usual dosages seem to have little effect but high concentration steroid mouthwashes may be effective in aborting the lesions. Intralesional injections of triamcinolone suspensions have been advocated, but there is little evidence of their effectiveness. So far as the non-erosive form of lichen planus is concerned, there is little evidence that any treatment is of value in influencing the duration of the lesions.

A recent form of therapy, introduced for the treatment of psoriasis, has been used with some success in the treatment of generalized lichen planus. This is psoralen photo-chemotherapy (PUVA). The drug 8-methoxypsoralen (given systemically) is activated by ultraviolet light to produce a thymidine antagonist in the skin. There are early reports of the adaptation of this technique to treat resistant and severe oral lichen planus, but it is by no means an established method of treatment. There have been discouraging reports on the use of retinoids (also normally used for severe psoriasis) in oral lichen planus.

## Pemphigus

Pemphigus is a disease of skin and mucous membranes in which bullae are produced as a result of acantholysis—the breakdown of the intracellular connections in the stratum spinosum of the epithelium. The bullae produced are, therefore, intra-epithelial and lie above the basal layer of the epithelium (Fig. 1.10). The patients are usually between the ages of 40 and 60 years at first diagnosis and there is an even sex distribution. There is a racial factor involved in that there is a high incidence of the condition among people of Jewish origin, although patients are by no means restricted to this group. The oral mucous membrane is involved in a high proportion of the patients and, in fact, half of all initial lesions are found in the mouth. Before the introduction of steroids the prognosis in this disease was very poor, the average time of survival was only 14 months from the onset of lesions. However, the introduction of treatment by systemic steroids has made the outlook considerably less gloomy although pemphigus must still be considered a very serious condition. It is, therefore, particularly important that the oral lesions should be recognized early in order that treatment may be started at the first possible moment.

The clinical picture in pemphigus is of widespread bulla formation in the skin and mucous membranes. The bullae, even of the skin, are fragile and break down rapidly to form crusted lesions (Fig. 8.7). The lesions of mucous membranes are even more fragile and rapidly break down with the formation of irregular ulcers, often with a ragged edge as a result of the split and fragile epithelium. The oral lesions may occur in any site within the mouth and oropharynx (Fig. 8.8) and may be accompanied by similar lesions of other mucosae; the vagina is often involved in this way. A number of clinical variants are known, the most clearly recognized being pemphigus vegitans in which the lesions formed by the rupture of the bullae are accompanied by the formation of exuberant granulation tissue: the 'vegitations' of the nomenclature.

In view of the rather nondescript nature of the oral lesions diagnosis may

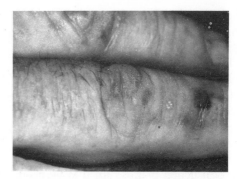

Fig. 8.7. Pemphigus. Intact and ruptured bullae of the fingers.

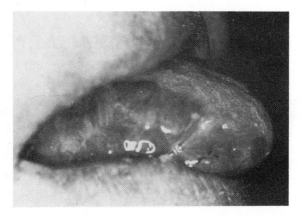

Fig. 8.8. Pemphigus. A large bulla on the tongue.

prove to be something of a problem. Incisional or excisional biopsy of the lesion is difficult since the epithelium is readily detached from the underlying tissues. A more uniformly successful diagnostic measure is the search for characteristic Tzank cells (cells of the stratum spinosum modified by the acantholytic process) in bulla fluid or in scrapings from the edge of a ruptured lesion (Fig. 1.12). The fragility of the epithelium may be recognized in some patients by the presence of a positive Nikolski sign: the epithelium of the skin or oral mucosa may be detached by sideways pressure. More recent developments, however, have made immunological methods the most important in the diagnosis of pemphigus. Circulating antibodies to the intercellular areas of the stratum spinosum may be demonstrated in over 90 per cent of patients and can be quantified to provide an indication of the progress of the disease. These antibodies are of the IgG group. Passive transfer between human patients and experimental animals with the production of lesions has been demonstrated and experimental acantholysis has been induced by IgG from affected patients. It would seem, therefore, that these autoantibodies represent an aetiological factor in the disease rather than a consequence of it.

Direct immunofluorescence demonstrates IgG class antibodies binding the intercellular substances and cell membranes in the stratum spinosum of affected epithelia, strongly so in and near the lesions, and rather less strongly at a distance from the lesions. These immune markers are highly specific for pemphigus, they are otherwise found only in patients with a history of severe burns (Fig. 8.9).

A rare hereditary form of pemphigus is known as benign familial chronic (or Hailey–Hailey disease). In this condition the histology of the lesions closely resembles those of pemphigus proper, with marked acantholysis. However, this is a much less aggressive disease than pemphigus and runs a protracted chronic course. The oral lesions closely resemble those of pemphigus and the rare occurrence of this variant should be borne in mind in differential diagnosis.

The treatment of pemphigus is a multidisciplinary matter; for example, an ophthalmologist may well be called in to help with the management of eye lesions. The particular role of the dentist and of the discipline of oral medicine is, first of all, to enable early diagnosis by the recognition of the oral lesions and, later, to help in the management of the often very severe oral lesions. The prognosis has been completely altered by the introduction of systemic steroids and many patients may lead reasonable lives maintained on substantial doses of prednisolone or some other steroid. Very high dosages are used initially to suppress bulla formation (of the order of 100 mg prednisolone daily), but this may often be slowly reduced to a maintenance dose of 15 mg daily or thereabouts. This systemic steroid therapy may well be supplemented by high concentration steroid-antibiotic mouthwashes since the oral mucosa is often less

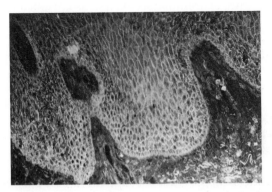

Fig. 8.9. Immunofluorescent demonstration of IgG group antibody deposits around the cells of the prickle cell layer of the epithelium in pemphigus.

responsive to treatment than the skin. Just as the oral mucosa seems to be particularly prone to attack by the antibodies early in the course of the disease, it also often remains quite severely affected when the skin lesions are in remission. This may be the case even when the level of the antibody titre is unmeasurably low. Very long term local steroid-antibiotic therapy may be called for, supplemented by such measures as antifungal therapy and the use of anaesthetic mouthwashes as found necessary. Oral hygiene may present a great problem and care must be taken if the teeth are not to be lost, an important dental factor, since the wearing of dentures may be difficult if not impossible.

## Pemphigoid

There are two basic clinical types of this condition: the first predominantly affecting the skin, the second the mucous membranes with only occasional skin involvement. Various terms have been used to describe these two main groups, referred to here as generalized pemphigoid and mucosal pemphigoid, respectively. The term 'generalized pemphigoid' is effectively synonymous with that often used by dermatologist: 'bullous pemphigoid'.

### Generalized pemphigoid

Generalized pemphigoid is a bullous disease which was for long thought to be a variant of pemphigus. However, it is now realized that the bullae are entirely subepidermal and are produced by the lifting off of intact epithelium; there is no acantholysis involved (Fig. 1.11). The patients with pemphigoid are, in the main, elderly, most being over the age of 60 years, although some very few may be considerably younger. As in pemphigus the sex distribution is even. There is no racial factor involved.

The condition starts with a rash, usually on the limbs, which becomes bullous only after an interval of several weeks or even months. When the bullae appear, however, they become quickly established over the whole of the body. The characteristic bullae are tense and much tougher than those of pemphigus; because of the absence of acantholysis the lesions are much less likely to break down immediately after formation. Mucous membrane lesions are confined to the mouth and occur only in some 20 per cent of patients. As in the skin, the oral lesions are relatively tough and remain either as intact or perforated bullae for a much longer period than those of pemphigus. This is not such a life-endangering disease as pemphigus and may run a chronic course over many years. However, before the introduction of steroids as a treatment, a considerable proportion of the patients died as a result of secondary effects such as intercurrent infection. With the use of systemic steroids most patients can be kept in reasonable comfort and, in some cases, it may be necessary to give only a single course (carefully tailed off) in order to induce a long period of remission.

Since the oral lesions appear rather later in the disease than the corresponding lesions of pemphigus, diagnosis is much less likely to be made on the basis of the oral lesions alone. However, diagnosis by excision of a mucosal lesion for examination may be a practical possibility although, even in these relatively tough bullae it is all too easy to completely loose the epithelial component of the biopsy specimen. Since the bullae are subepithelial, acantholytic Tzank cells are not seen in the aspirated bulla fluid; this can provide a valuable test in the differential diagnosis of the bullae. Circulating autoantibodies can be detected by immunological investigation and these can be shown by immunofluorescent techniques to bind onto the basement membrane of the epithelium (rather than onto the cells of the stratum spinosum as in pemphigus). These are found in about 70 per cent of patients. Direct immunofluorescence on biopsy material shows IgG antibodies in a linear distribution along the basal zone (Fig. 8.10). IgA, C3, or C4 may also be present, although IgA deposition in the absence of IgG is characteristic of another skin disease, dermatitis herpeteformis.

## Mucosal pemphigoid

In mucosal pemphigoid oral lesions are almost always present; other mucosal lesions are relatively common, but skin lesions are few and infrequent. Just as in the oral lesions of generalized pemphigoid, described above, the bullae are relatively tough and likely to be recognizable as blisters when first seen (Fig. 8.11). However, this is not invariably the case: some patients present with eroded areas of the mucosa which are only diagnosed on biopsy. In most dermatology texts scar formation is described as a characteristic of oral pemphigoid. In the author's experience this is by no means invariably so although it does most certainly ocur in a number of cases and is a very important factor. Indeed, the significance attached to scarring, with its particular importance in terms of the eyes, had led to the use of the term 'cicatricial pemphigoid' as being almost synonymous with mucosal pemphigoid. In the case of the author's patients this is certainly not so but the significant group of patients in which this is the case is discussed below.

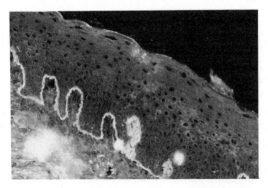

**Fig. 8.10.** Immunofluorescent demonstration of IgG group antibody deposits along the basal zone in pemphigoid.

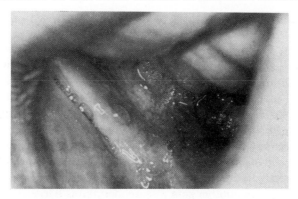

**Fig. 8.11.** Mucosal pemphigoid. An intact bulla on the alveolar mucosa. In this patient the majority of bullae were in the typical site, the soft palate.

The clincial picture in mucosal pemphigoid is widely variable. The patients vary from under thirty years of age to over seventy at first diagnosis with a preponderance of the 50–70-year-old group. There is a 4:1 preponderance of female patients. Many patients have bullae occuring on the oral mucosa only, with no skin or other mucosal involvement. The frequency and extent of bulla formation varies from virtually continuous, and multiple to intermittent and single. The bullae, when established, are in general painless, but there may be discomfort when the bulla if forming and after rupture. A smaller subgroup of patients are clearly recognizable. These are the ones most often described under the headings of benign mucous membrane pemphigoid or cicatricial pemphigoid. The lesions are, histologically and immunologically, identical with those in other forms of mucosal pemphigoid. However, there is a very characteristic site for the oral lesions, in virtually all cases the initial lesions are of the soft palate and scars may be left following rupture and healing of the bullae. Most patients are female and over the age of 60 years. In a considerable proportion of patients (as many as 75 per cent) the conjunctiva is eventually involved and, since the lesions heal with scarring, vision may be affected. The definition of the term 'desquamative gingivitis' has been discussed in Chapter 6 and referred to above in the account of gingival lichen planus. As has been pointed out, a number of patients with this clinical picture are shown on biopsy to have mucosal pemphigoid. On the whole this pattern of the disease is relatively self-limiting, but in a few patients the gingival lesions may extend, become more evidently bullous and be very difficult to treat.

In mucosal pemphigoid the immunological changes are similar to those in generalized pemphigoid, although circulating antibodies can be demonstrated far less frequently. Direct immunofluorescence is also less reliable and currently proves positive in something of the order of 75 per cent of histologically proved cases although the sensitivity of the technique is increasing with improved reagents.

Treatment of this condition is on a local basis. Systemic steroids are rarely indicated, but betamethasone-17-valerate in spray form is both particularly effective and also convenient for application to the soft palate. Symptomatic treatment for the relief of oral erosions caused by the perforation of the bullae may also be necessary. A few patients with recurrent multiple bullae need long-term steroid-antibiotic mouthwashes. Often the control, even with high concentrations of steroids, is not particularly good. At the other extreme, in the case of intermittent mucosal pemphigoid where the bullae may occur only once every few months, it is difficult to suggest any action other than the use of antiseptic mouthwashes to relieve discomfort when the bulla ruptures.

An important facet of the management of patients with mucosal pemphigoid is that of the avoidance of eye complications; conjunctival scarring can be so severe as to lead to eventual blindness. Thus, all patients with the biopsy proved condition should have the benefit of an ophthalmological opinion.

It should perhaps be added that the concept of rigid classifications of bullous diseases, clearly defined on clinical and histological grounds, is to some extent giving way to the idea that transitional conditions may occur. This recognition by dermatologists of a spectrum of presentation of bullous skin disease is paralleled by similar observations in the case of oral lesions. Although the great majority of these may be classified with some degree of confidence into one or other of the accepted disease patterns, there remain some lesions which both clinically and histologically seem equivocal.

## Oral blood blisters

Some patients develop spontaneous blood filled bullae ('blood blisters') of the oral mucosa from time to time. These have been described under the unfortunate title 'angina bullosa haemorrhagica'. The usual pattern is that the patient feels a sharp pricking sensation in the mouth (most usually on the palate) and finds that a blood filled blister has suddenly developed. This most commonly occurs when the patient is eating. The bullae may be quite large (up to 2–3 cm in diameter) and the patient may be in considerable fear of obstruction. The blister eventually ruptures or is perforated by the patient, and healing occurs uneventfully.

These patients have no demonstrable abnormality of the blood clotting mechanism, although patients with thrombocytopoenia may also develop blood blisters. Both male and female patients have been described, over a wide age range. The method of formation of the blisters is not known; it is speculated that the basic mechanism is of bleeding from the capilliary bed below a basal zone which is for some reason weakened. This may indeed be the mechanism, but the reasons for it are far from clear.

Management consists of perforating the blister to release the contents if an intact blister is seen. Often, however, only a ruptured bulla is presented for examination, the patient having perforated it as a first measure. No preventive treatment is known.

## Erythema multiforme

Erythema multiforme is an acute disease of skin and mucous membranes with a wide range of clinical presentations, hence the term 'multiforme'. It may be precipitated by a wide range of stimulating factors including infections (particularly viral), neoplasia, pregnancy, and drugs of various kinds, but in about half of the cases no such inducing factor is found. In the most fully developed form (alternatively known as the Stevens–Johnson syndrome) there is widespread involvement of the skin and mucous membranes, but in the more usual restricted form the oral mucous membrane is mainly involved, with no more than minor lesions in other sites. The patients are predominantly young adults, males being affected three times more frequently than females.

The main feature of an attack is the sudden development of widespread erosions of the oral mucosa, characteristically involving the lips. The erosions are produced by the disintegration of subepidermal bullae, lesions which only rarely last long enough to become a diagnostic feature. The erosions on the lips (especially the lower lip) are accompanied by crusting and bleeding and are, if not absolutely diagnostic, strong pointers to the nature of the condition (Fig. 8.12). There is often a cervical lymphadenitis and pyrexia, and the patient feels ill and depressed. The accompanying skin lesions, when present, have a characteristic target appearance which is diagnostic. An attack gradually subsides after some 10 days, but there is a considerable likelihood of recurrence after a variable period from no more than a few weeks to a year or so. However, the subsequent attacks become gradually less intense and, over a total period of 2 or 3 years, the condition gradually subsides. In those cases precipitated by viral infection the attack may follow a lesion of recurrent herpes, the interval

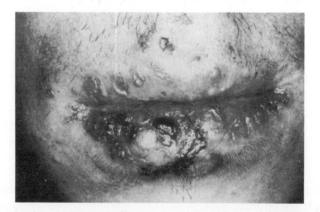

Fig. 8.12. Erythema multiforme, showing typical lesions of the lips.

usually being of the order of 10 days or so. In a few patients the attacks may be very closely spaced or even continuous without any evident precipitating cause.

The initial diagnosis is entirely clinical, the important differential diagnosis being from a primary herpetic stomatitis (Fig. 8.13). In the case of a recurrent attack, however, herpetic stomatitis may be confidently excluded since this is an isolated event. Similarly, a history of recurrent herpes simplex also excludes the possibility of herpetic stomatitis. The involvement of the lips is a strong indication of the diagnosis of erythema multiforme and the presence of target lesions of the skin can be taken as almost conclusive evidence for the diagnosis. Biopsy of the oral lesions shows a rather non-specific histological picture with degenerative changes in the epithelium and sub-epithelial bulla formation. Because of the acute clinical features, however, the diagnosis does not often depend on the histological appearance of the lesions.

Treatment of the cases restricted to the mouth depends on the use of local steroids to which there is usually a rapid response. A triamcinolone-chlortetra-cycline mouthwash (or its equivalent) with a moderate concentration of the steroid is likely to give symptomatic relief and to effectively reverse the process in a few days. In the more severe cases, however, particularly when other mucous membranes are involved, when the skin lesions are severe, or when the eyes are affected, a short course of systemic steroids may be necessary to abort the attack. A course of 40 mg prednisolone daily for 1 week, rapidly reducing over a further week, is used together with a systemic antibiotic to reduce the risk of secondary infection. Such patients are likely to be sufficiently ill to require hospital admission. In those patients in whom the attacks are precipitated by recurrences of facial herpes, treatment of the herpetic lesion at the earliest possible moment

**Fig. 8.13.** Erythema multiforme affecting the tongue and closely resembling the picture in acute herpetic stomatitis.

with either 5 per cent idoxuridine or acyclovir may lead to a marked reduction in severity or even abolition of the expected attack of erythema multiforme.

## Epidermolysis bullosa

Epidermolysis bullosa is a condition (or rather a group of related conditions) which is very unlikely to be diagnosed on the basis of oral symptoms alone but which may be of great significance to the dental surgeon in his treatment of an affected patient. The predominating feature is extreme fragility of the epithelium, the result of bulla formation occurring either spontaneously or as a response to minimal degrees of trauma. In the 'simplex' form of the disease there is a minimal bulla formation and the mucous membranes are rarely involved. At the opposite extreme of activity is the 'lethalis' form in which the affected child at birth presents with an extreme degree of fragility of the skin and mucous membranes which is incompatible with life. Few subjects with the lethalis form of epidermolysis bullosa survive for more than a few weeks and many die *in utero* or at birth. In the dystrophic form of the disease, however, the patients may, with care, survive for several decades in spite of an incapacitating degree of skin and mucosal involvement, and it is this group of patients in whom dental management is fraught with difficulties.

Dystrophic epidermolysis bullosa is inherited as a recessive trait and, in affected individuals, bullae are seen at, or soon after, birth. The lesions are produced in response to the most minor degrees of trauma and eventually heal with scar formation, this leading eventually to gross tissue deformity, particularly of the extremities. The oral mucous membrane is equally susceptible to damage and, as a result of the repeated scar formation, opening of the mouth may become greatly restricted and the tongue bound down. The essential pathology in this form of the disease is dermal; the fragility of the tissues is due to deficient collagen formation in the subepithelial structures. This situation is paralleled in the teeth, there being abnormalities in dentine formation which lead to hypoplasia and a high susceptibility to caries. Since conservative dental treatment or even effective oral hygiene measures may be almost impossible in these patients, the resulting grossly carious teeth, associated with a restricted access and extreme mucosal fragility, present a major problem to the dental surgeon. It has also been reported that the oral scars have, in some patients, been followed by the onset of leukoplakia and, finally, carcinoma, but this cannot be regarded as a characteristic of the disease.

Treatment of epidermolysis bullosa is largely symptomatic. High doses of steroids may be used to suppress the lesions, but in infancy these may cause damage to the growth centres. Systemic antibiotics are freely used to treat the secondary infections which frequently occur and antibiotic mouthwashes may be used in the control of any particularly severe episode of oral ulceration.

It should perhaps be mentioned that the term 'oral-acquired non-dystrophic epidermolysis bullosa' has recently been used to describe a localized condition in which bullae form on the gingivae in response to trauma. There is no apparent connection with epidermolysis bullosa as it is generally understood and the term has been used descriptively because of the fragility of the epithelium. Little is as yet known of this condition.

## Other dermatoses

Oral lesions have occasionally been reported in a wide range of dermatological diseases other than those described. For instance, psoriasis has previously been mentioned as being such a condition and dermatitis herpetiformis is another in which the available information is insufficient to describe a characteristic oral lesion. There may also be some confusion, in these less well known conditions, between the oral lesions proper and the secondary effects of treatment. For instance, in the case of dermatitis herpetiformis, two types of oral lesions have been described, one being vesicular and one resembling lichen planus. It may well be that this second form of lesion represents a reactive condition consequent to dapsone therapy for the condition. However, this is by no means proven and, in the case of such a patient investigated by the present author, the lichenoid form of lesion was found to bear little histological resemblance to that of lichen planus. This is one example of the occasional difficulty experienced in the diagnosis of the more unusual oral lesions in skin disease.

## Connective tissue diseases

This is a compendium term used to describe a number of diseases with a similar, but by no means identical immunological background. They are not skin diseases, but there are skin lesions in a number of them and it is common practice to group them with the skin diseases for descriptive purposes. The group includes rheumatoid arthritis and Sjögren's syndrome (in which there are no skin lesions at all). These are discussed in Chapters 5 and 12. Apart from these two conditions lupus erythematosus has important oral implications as has mixed connective tissue disease and scleroderma, although these last two are relative rarities.

## Lupus erythematosus

The group of diseases included under this heading present with a wide range of symptoms, but all result from abnormalities of the connective tissues brought about by an autoimmune process. Two main clinical entities are recognized although there are many variations. These are systemic lupus erythematosus and chronic discoid lupus. In systemic LE there are widespread changes in the connective tissues with secondary effects in the cardiovascular, musculoskeletal, and other systems as well as in the skin. The course of the disease varies from a relatively mild chronic condition to a rapidly fatal process and an equally wide range of skin reactions may occur. These are paralleled by an equally diverse range of oral symptoms, the most commonly described being superficial erosions and erythematous patches on the mucosa. It would seem very unlikely that the initial diagnosis of the disease would be made on the grounds of oral lesions alone, but the possibility of such an aetiology for unrecognized oral lesions should be borne in mind. In particular, it should be remembered that most clinical descriptions of the lesions of all lupus variants resemble those of lichen planus. Histologically, also, there is a close resemblance between these conditions. The final diagnosis of systemic LE is likely to be made as a result of the immune abnormalities present and by the recognition of the characteristic cells (LE cells) formed when the leukocytes from the patient are incubated with the patient's own serum. The LE cells are polymorphs which have ingested degenerate nuclear material in the presence of a specific antibody contained in serum from an affected patient (the LE factor). Over 90 per cent of patients have circulating anti nuclear antibodies. This is the most significant immunological screening test. If there are skin lesions the histology and immunofluorescent findings on biopsy are as in chronic discoid lupus (below).

Just as a lichenoid reaction may occur as a response to drugs, systematic LE may be precipitated in the same way and by an equally wide range of drugs. Hydralazine, used in the management of refractory hypertension, is the most quoted example, but most of the drugs mentioned above as implicated in lichenoid reactions have also been reported as causing lupoid reactions.

Treatment of systemic LE is essentially with steroids, often in high doses and with the addition of steroid sparing drugs such as azathioprine. The oral lesions may be both very painful and difficult to treat. High concentration steroid-antibiotic mouthwashes are most useful, together with such supporting measures as lignocaine mouthwashes.

Chronic discoid lupus is a much more restricted form of the disease which presents largely as a skin disorder and without the widespread generalized abnormalities found in the systemic form. The skin lesions, which result from degenerative changes in the subepithelial connective tissues, present as scaly red patches which later heal with scar formation. The face is the area most commonly affected and there is often a symmetrical distribution of the lesions over the nose and cheeks, the so-called butterfly pattern. The patients are predominantly female (F:M = 2:1), the age of incidence being widely distributed, but having a peak in the fourth decade of life. The first appearance of the lesions may follow some form of trauma (such as an unusual degree of exposure to sunlight) and later exacerbations may follow repeated trauma of this kind. This is not a particularly rare condition; the incidence in an English dermatological clinic has been assessed as 0.4 or 0.5 per cent of all new attendances. Oral lesions are found in a considerable proportion of patients with chronic discoid lupus, although the estimated incidence varies very widely from 3 to 50 per cent according to the source quoted. Although lesions may be found on any part of the oral mucosa the characteristic site is on the lips. The lesion starts as an area of erythema which develops to a thickened, rather crusted, lesion with a white margin. The histological appearances are of epithelial atrophy at the centre of the lesion with hyperkeratosis at the margins with a close resemblance to the changes in lichen planus. The fundamental difference between the histological findings in lichen planus and in LE (and other connective tissue diseases with skin and mucosal lesions) is that the subepithelial band of lymphocytes, relatively evenly distributed in lichen planus, has a tendency to a follicular distribution in LE. Direct immunofluorescence in LE gives variable results with homogeneous or granular deposits of IgG, sometimes with IgM and complement components, at or below the basal zone.

Treatment of chronic discoid LE is often symptomatic with the use of steroids to suppress the lesions. Parenteral treatment, oddly enough, is with the antimalarial drugs which may completely suppress the symptoms but which may also introduce a wide range of side effects, some minor and some more serious; for example, the production of corneal deposits and retinopathy. The most significant consideration, so far as the oral lesions are concerned, is the possibility of malignant change. It is difficult to assess the incidence of this from the published figures, but there is no doubt that cases involving malignant transformation in lip lesions have been documented. It is, therefore, necessary to observe the lesions on a long-term basis.

## Other connective tissue diseases

Mixed connective tissue disease is a condition in which a number of the characteristics of other diseases in the group are found in a single patient. It is a rare condition, but of interest in the present context since there are two possible oral diagnostic indicators of the disease. The first is a lichen planus like lesion of the oral mucosa, but with a histological appearance resembling that of LE. The second is the involvement of the trigeminal nerve in the neurological changes which may occur. This may lead to facial anaesthesia. The most important immunological indicator in this condition is the presence of anti nuclear antibodies of the speckled type.

In scleroderma (systemic sclerosis) there is widespread fibrosis of the subepithelial structures in the skin and gut together with other organs. There may also be other elements of connective tissue disease present. The oral problem is largely in the constriction of the tissues which causes increasing lack of ability to open the mouth with the inevitable results of poor oral hygiene coupled with great difficulty in carrying out dental treatment.

## Gastro-intestinal disease

Oral lesions may occur in a number of lower gastro-intestinal tract diseases. In a few, these are primary oral lesions resembling those of the lower gut, but in almost all, secondary lesions may be induced by factors such as malabsorption. Not only the disease process, but also the elimination of parts of the gut by surgery may result in these secondary changes.

Some diseases of the gastro-intestinal tract are of particular significance in the field of oral medicine and are discussed below.

## Coeliac disease

Coeliac disease has been recognized in children since 1888. When the affected children are weaned onto a solid diet they develop diarrhoea, vomiting, loss of appetite, and wasting together with secondary effects of malabsorption such as anaemia. A similar (but not identical) disease known as tropical sprue has been known in adults for a long time and it is interesting that the name 'sprue' was derived from the Dutch word for aphthous ulcer, this indicating the high proportion of sufferers with oral ulceration. However, it has been since recognized that adult coeliac disease may occur in the non-tropical environment and although the symptoms are similar to those of sprue, it is considered that these are different diseases.

Both children and adults with coeliac disease are found to have an abnormal jejunal mucosa, the villi normally found being partially or completely absent. Apart from this flattening of the mucosa there is an inflammatory cell infiltration of the lamina propria and lengthening of the crypts. The mechanism which leads to these changes and also to the production of symptoms is not clearly defined, but there is no doubt that they are associated with some form of abnormal reaction to gluten which is a natural constituent of the diet, present particularly in wheat products. If gluten is eliminated from the diet the symptoms resolve and after a time the jejunal mucosa reverts to normal. If gluten is then reintroduced the mucosa again becomes flat and the symptoms recommence. The amount of gluten required to precipitate symptoms in this way may be small.

The nature of this reaction to gluten is not known although recent work has implied an interaction between a gluten constituent and the reticulin fibres of the jejunal submucosa. The immunological data is confused and suggested IgA abnormalities have not been confirmed. There is some degree of association with the HLA B8 locus. As yet, reliable tests for antibodies to α gliaden in the serum have not become widely available although there is a reasonably specific skin test for hypersensitivity to gluten. The issue is further complicated by the fact that some patients may show a positive skin reaction to gluten whilst having a normal jejunal mucosa.

The gut changes in adults may produce less marked symptoms than in children and may pass unrecognized. However, most affected adults have some degree of malabsorption which may result in a macrocytic anaemia, and in very variable degrees of gut upset and weight loss. An itchy skin rash or a more fully developed dermatitis herpetiformis may also be present. The greater significance in the context of oral medicine is the occurrence of recurrent oral ulceration in some of these patients. This was pointed out by Truelove and Reynell who remarked that the correct treatment for such oral ulceration was a gluten-free diet. However, it has only recently become clear that a substantial number of patients with recurrent oral ulceration, particularly of the minor aphthous or herpetiform type, show haematological evidence of malabsorption on screening and of these a quite high proportion show a flattened mucosa on jejunal biopsy. In such patients the adoption of a gluten-free diet results in the elimination of the oral ulceration and often in subjective feelings of better health. The initial studies of this relationship implied that the proportion of these patients was of the order of 25 per cent of those with ROU. However, it has since become clear that the figure is much lower, of the order of 6 per cent of those screened. There is a similar, well established relationship between coeliac disease and the skin disease dermatitis herpetiformis, although in this case most patients with dermatitis herpetiformis are found to have a flat jejunal mucosa.

It is a surprising fact that all patients reported to have an otherwise unsuspected flat jejunal mucosa are adults. No patient under the age of 16 has as yet been shown to have this relationship between a flat jejunal mucosa and ROU in the absence of overt gastro-intestinal symptoms or other evidence of coeliac disease. It is not known at what age the adult pattern of the association may first occur.

It is generally accepted that patients with ROU should not be submitted to jejunal biopsy unless there is evidence of malabsorption or some other indication of gastrointestinal abnormality. In view of the fact that children are not shown to have this particular form of association it is felt that the indications for jejunal biopsy in childhood depend largely on the degree of systemic disturbance. It has been shown that, even in children with ROU and recognizable haematological abnormalities, the origin of these in the absence of other symptoms is unlikely to be coeliac disease. In adults with ROU the haematological indications for jejunal biopsy are a reduction in haemoglobin, serum iron, serum or red cell folate, or B12 levels. A full haematology screen is essential—a simple blood count is quite useless for this purpose.

The diagnosis of coeliac disease has further importance since the affected patients are known to have an increased incidence of neoplasia of the gut. It is not as yet known whether the adoption of a gluten-free diet reverses this tendency.

## Crohn's disease

Crohn's disease was first described in 1932 as regional ileitis. The original paper included a description of mucosal inflammation and occasional ulceration of the affected gut, lymph node hyperplasia leading to obstructive oedema, and the production of granulomatous lesions. Shortly after the first description of the disease it became clear that it was not of necessity confined to the ileum and lesions have since been described throughout the whole of the gastro-intestinal tract. The clinical progress of the disease is very variable with inactive phases. However, in the more aggressive stages of the disease, there may be pain, diarrhoea, and malaise; fissure formation in the affected gut may progress to the production of fistulae. There may be arthropathies and skin granulomas associated with the disease as well as problems arising from malabsorption. The aetiology of the disease is not known and the eponymous description has, therefore, been maintained. Treatment is medical, using steroids, azathioprine, sulphonamides, and replacement therapy to correct malabsorption. In some advanced cases surgery is necessary.

The oral lesions in Crohn's disease parallel those in the gut. These consist of lesions of the buccal and labial mucosa, which may be thrown into corrugations or may show marked proliferation (Fig. 8.14). These lesions may be primarily or traumatically ulcerated, although this is not usual. There are palpable lymph nodes in the submandibular and submental areas with oedema of the lips and, occasionally, cheeks (Fig. 8.15). Angular cheilitis is common and there are occasional superficial fissures of the swollen lips. A characteristic stippled gingivitis is present in many of the patients, affecting the attached gingivae. Of these lesions it is the swelling of the lips which often brings the condition to notice; in more advanced cases this may become quite gross (Fig. 8.16).

Definitive diagnosis depends on the presence of characteristic granulomas in the buccal lesions. These have a non-caseating tuberculoid structure, often with giant cells present. These vary greatly in number and distribution; in some lesions there are large numbers, in others few. In some cases the granulomas are only found deep in the tissues. There are, therefore, indications for a relatively deep biopsy followed by a microscopic search through all the tissue if they are not at first seen. Biopsy of the lips is quite unproductive—the cause of the oedema is lymphatic obstruction, often at a distance, and biopsy tissue shows no characteristic changes. Biopsy of the gingivae also demonstrates the presence of granulomas, although this site is not recommended as a first choice for biopsy for reasons given in Chapter 6.

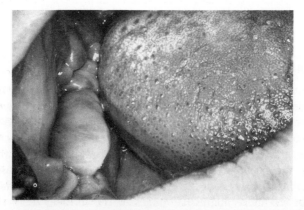

Fig. 8.14. Crohn's disease. Florid granulomas of the buccal mucosa.

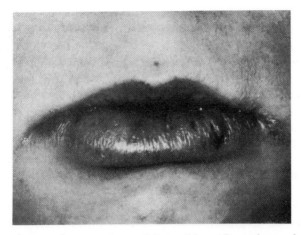

Fig. 8.15. Crohn's disease. Swollen and fissured lower lip with angular cheilitis.

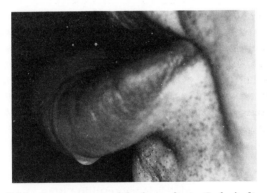

Fig. 8.16. Gross oedema of the lower lip in Crohn's disease.

It is quite clear that in an increasing number of patients these oral lesions represent the first stages of systemic Crohn's disease and that lesions elsewhere in the GI tract may follow, often in an aggressive form. It is also abundantly clear that, even when initial full studies show no GI tract abnormality in these patients, this situation may not continue and that later investigations may show abnormalities developing. In these circumstances the statement made in Chapter 4 is reiterated: that these patients should be considered as suffering from oral Crohn's disease until proved otherwise. This being so, initial diagnosis of the oral lesions should be followed by a full haematological study to detect malabsorption together with contrast radiological studies of the GI tract. In view of the fact that entirely unsuspected indications of generalized Crohn's disease have been found in a number of such patients radiography would seem an essential step in the first investigation.

Treatment of the oedematous lips is unsatisfactory. In spite of statements to the contrary neither systemic nor intralesional steroids have been found by the author to exert any lasting improvement, although transient reduction in swelling may be obtained. Local steroid creams may help to a minimal extent. On the other hand, the angular cheilitis (and often lip fissures) respond rapidly to 1 per cent hydrocortisone cream. Occasionally, ulceration of the buccal lesions is severe and a short course of systemic prednisolone may be very helpful. In general, however, ulceration is minimal and only antiseptic mouthwashes are required to minimize discomfort.

It must be reiterated that the most important factor in the management of these patients is an awareness of the possibility of the development of lower gut lesions. The co-operation of a physician with a special interest in gastroentero-logy is essential.

## Ulcerative colitis

Ulcerative colitis is a disease, usually beginning in young adult life, in which inflammatory changes in the colonic mucosa and submucosa leads to wide-spread ulceration. This ulceration may be complicated by haemorrhage, perfora-tion, and occasionally by the eventual onset of malignancy. Pain, diarrhoea, and generalized abdominal discomfort are the predominant symptoms. The aetiology of ulcerative colitis is not clear although it seems to be one of the 'diseases of civilization'. Immunological studies have suggested an autoimmune process possibly including a cross-reactivity with bacterial antigens. The oral lesions associated with ulcerative colitis have been described as 'severe aphthae' and candidiasis has also been described. It may well be that these represent the secondary effects of nutritional deficiency resulting from malabsorption.

However, in a few patients characteristic skin lesions are produced (deep ulcers with undermined edges); the condition is known as pyoderma gangreno-sum and in these circumstances oral ulcers of similar characteristics may also occur. The term pyostomatitis gangrenosum has been suggested to describe these destructive lesions. A less aggressive skin lesion has been described as pyoderma vegetans and has been demonstrated to have an oral equivalent, pyostomatitis vegetans. In these lesions a number of small microabscesses are present in a rather granular lesion of skin or mucosa. These lesions have all been described in patients with other forms of colonic pathology than classical ulcerative colitis. It has been suggested that they are the result of circulating immune complexes which cause necrosis.

The treatment of ulcerative colitis is by the use of sulphasalazine (a sulphono-mide which seems to have a specific effect on this condition) and by steroids, used either locally (as pessaries or enemas) or systemically. Surgery may be necessary if medical treatment fails. The skin and oral lesions might be expected to respond to the treatment for the generalized condition, this is likely to be intensive since the lesions appear during active phases of the disease. Supporting measures, such as the use of antibiotic mouthwashes, may be helpful, as in other forms of oral ulceration.

# 9

# Blood and nutrition: endocrine changes and drug reactions

## Blood and nutrition

It is well known that lesions of the oral mucosa may occur in patients with abnormalities of the blood, in particular the appearance of such symptoms as a sore tongue or angular cheilitis in anaemic patients has often been described. However, it has more recently become recognized that such oral symptoms may be the result of relatively minor changes in the condition of the blood and that they may occur early in the disease process, even before abnormalities can be demonstrated by a simple blood examination. Thus, an early diagnosis of the blood disorder may depend on a recognition of the significance of the oral symptoms. The great majority of these patients are suffering from anaemias of various kinds and, hence, the major interest is centred on this group of conditions but it must be borne in mind that abnormalities of the white cell and platelet components of the blood may also be reflected in oral changes.

## Anaemias

The red cells (erythrocytes) are the structures predominantly concerned with the transport of oxygen to the tissues by means of the iron-containing substance haemoglobin within them. They are normally regular, biconcave discs (Fig. 9.1), but if disturbances of formation occur they may become quite irregular in size and shape. Such irregularity is often a sign of impaired function. The formation of the erythrocytes within the bone marrow is stimulated by a number of nutritional factors, the two of greatest significance being vitamin $B_{12}$ and folic acid. Both of these substances are absorbed from the gut and must present in balanced quantities for normal red cell production to take place, even when sufficient iron is available for the synthesis of haemoglobin. Absorption depends on a normal gut mucosa and, in particular, on the presence of the so-called intrinsic factor which is synthesized in the gastric mucosa and which must be present before absorbtion of vitamin $B_{12}$ can take place. If there are abnormalities which lead to failure of intrinsic factor synthesis, vitamin $B_{12}$ cannot be absorbed from the gut and must be replaced systemically. Lack of either vitamin $B_{12}$, folic acid, or intrinsic factor disturbs the absorption process and affects the marrow, and consequently, red cell production. The erythrocytes formed under

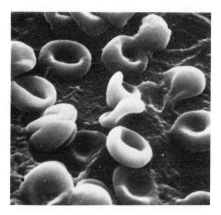

**Fig. 9.1.** Scanning electron micrograph of normal human erythrocytes.

these conditions are larger than normal and their function is severely disturbed. The resultant clinical conditions are known as megaloblastic anaemias. Pernicious anaemia is one of these, resulting from atrophy of the gastric mucosa and consequent lack of intrinsic factor. More complex malabsorption syndromes also occur, involving failure to absorb not only vitamin $B_{12}$, but also folic acid and iron compounds. The term megaloblastic anaemia refers to the change in size and structure of the basic marrow cell from which the erythrocytes are derived; the large circulating erythrocytes formed from these abnormal stem cells are macrocytes. Similar large circulating red cells may be found in other anaemias which are not dependent on abnormalities of the vitamin $B_{12}$/folic acid metabolism (for instance, in some iron deficiency anaemias) and such macrocytic anaemias form a separate group which, evidently, will not respond to treatment with vitamin $B_{12}$ or folic acid.

The situation is complicated by the fact that, in multiple deficiencies, the tendency to microcytosis as a result of iron deficiency may be counteracted by a tendency to macrocytosis caused, say, by folate deficiency. The result may be a normal MCV in a patient with both deficiencies present; in these circumstances a routine blood count will be returned as normal. It is also clearly established that patients with folate or $B_{12}$ deficiencies may well develop oral signs and symptoms before the erythrocytes are affected and before anaemia develops. Again, an argument for the necessity for a full haematological examination in these patients. It may well be that the patients developing oral signs at an early stage of a haematological abnormality represent a selected group with an unusually sensitive mucosal response to the changes.

## Iron deficiency

A much more common cause of anaemia than failure to absorb vitamin $B_{12}$ or folic acid is iron deficiency which leads to inadequate haemoglobin synthesis. The deficiency may be due either to inadequate intake of iron or to excessive blood loss as in menstrual abnormality or gastro-intestinal bleeding. In iron deficiency anaemia, the essential feature observed is a reduction in the haemoglobin concentration within the erythrocytes. They appear pale on microscopic examination, and there may be variations in size and shape. However, the number of erythrocytes per unit volume may not vary greatly from its normal value; the erythrocyte count is not considered to be a particularly important diagnostic test in most cases.

Of the total iron content of the blood, by far the greater proportion is combined in the form of haemoglobin within the red blood cells. A small fraction is present in the plasma, bound to a specific protein, transferin, and represents the transport iron made available from the body reserves to replace haemoglobin losses. If the stores become exhausted there is a period of latent iron deficiency in which the haemoglobin concentration is within normal limits and the erythrocytes of normal size and form, but the serum iron concentration is reduced. This is sideropenia, an iron deficiency which may affect the tissues and is capable of causing oral symptoms, but which does not produce anaemia since the haemoglobin remains unaffected. When the serum iron is depleted in this way, the degree of saturation of the transferin by the iron will evidently be reduced; this forms the basis of a valuable diagnostic test. In more complex conditions than iron deficiency, there may also be a reduction in the circulating transferin and so the degree of saturation may remain high in spite of a low serum iron value.

The binding protein for stored iron is ferritin. If the test is available, the serum ferritin level is a further good indicator of iron deficiency (Chapter 2).

The stages of iron deficiency may be summarized.

(1) Pre-latent iron deficiency, in which the body stores of iron are depleted, but the circulating haemoglobin and serum iron remain within normal limits.

(2) Latent iron deficienty, in which the body stores are exhausted and the serum iron reduced. The haemoglobin concentration remains unaffected.

(3) Iron deficiency anaemia in which the haemoglobin concentration is reduced. The serum iron will, evidently, also be low.

Oral symptoms may appear in the second and third of these stages.

## Haemolytic anaemia

There is a further group of anaemias, the haemolytic anaemias, in which the essential abnormality is an increase in the rate of erythrocyte destruction. Under normal conditions, the red cells last for about 100 days, but in haemolytic anaemias the life may be reduced to only a few days. Haemolytic anaemias may be due to an intrinsic defect or may be acquired, an important, although relatively uncommon, cause of the acquired form being the effect of some drugs.

Among the drugs which might be used in dentistry and which might very occasionally exert this effect are mefenamic acid (Ponstan), sulphonamides, phenacetin, and penicillin. However, the most important haemolytic anaemia in terms of dental practice is sickle-cell anaemia, although this condition is somewhat different from the others under consideration in that its major significance to the dentist is not in the production of oral lesions.

## Sickle-cell disease

The sickle-cell diseases are a group of genetically determined conditions in which the red cells contain an abnormal haemoglobin, HbS. When HbS looses oxygen it undergoes changes which produce distortion of the cells, the sickle effect. HbS is transmitted as a dominant characteristic in a high proportion of those with negro ancestry and in some families from Mediterranean countries and from the middle and far east. The carrier state (in which the patient inherits the condition from only one parent) is known as the sickle-cell trait and is by far the more common condition. In this trait the proportion of HbS in the red cells is low and sickling does not take place under normal circumstances. However, sickling may occur in conditions of low oxygen tension and, if this change does take place, the oxygen carrying capacity of the blood is greatly reduced with the consequent possibility of dangerous anoxia in the patient. The fully developed sickle-cell anaemia is the result of the donation of two HbS genes, one from each parent. In this case, the proportion of HbS is high and sickling occurs under normal body conditions. In these patients, the oxygen carrying capacity of the blood is poor, there is impairment of vascular flow and a haemolytic anaemia results from the shortened life of the abnormal erythrocytes. This is a much more severe condition than the sickle-cell trait and the symptoms of general ill-health are so marked as to make it very unlikely that a patient would present for dental treatment undiagnosed. However, there is no such guarantee in the case of the sickle-cell trait and the only way to identify these patients is to carry out a test for the presence of HbS. Fortunately, the initial screening test for this condition is relatively simple and is easily performed in the laboratory.

## Oral signs and symptoms in anaemias

A wide range of oral signs and symptoms may appear in anaemic patients as a result of basic changes in the metabolism of the oral epithelial cells which are particularly susceptible to minor variations in the quality of the blood supply. These changes give rise in their turn to abnormalities of cell structure and of the keratinization pattern of the oral epithelium, the end result often being atrophy. This atrophy seems particularly to affect the complex filiform papillae of the tongue which may be almost completely lost. However, the changes are by no means restricted to the tongue and ulceration or generalized soreness may occur over the whole of the oral mucosa. Apart from this type of symptom the patient with anaemia or latent iron deficiency is particularly susceptible to infection by

*Candida albicans* and angular cheilitis or thrush may occur. In addition to these changes, the patients may complain of disturbances of taste sensation. It has been suggested that this is due to atrophy of the tongue epithelium with resulting disturbance of the underlying nerve endings, but such a disturbance of taste has been observed in patients with apparently clinically normal tongue epithelium. A number of reports have described the results of the investigation of oral lesions in anaemic patients and it has become evident that there is no clear correlation between the oral symptoms and the basic aetiology. Sore tongue, taste disturbance, generalized stomatitis, ulceration, candidiasis, angular cheilitis, or gingivitis may occur in any of these patients.

The same considerations apply in relation to folate and $B_{12}$ deficiency. It has been pointed out above that very early deficiencies of either of these factors may result in oral mucosal changes; these are certainly not due to secondary anaemia since none may be present. The precise reason for these changes is not known. Work has been carried out on the role of $B_{12}$ in maintaining the stability of the mucosal epithelium and the results suggest a similar relationship as between $B_{12}$ availability and the production of normal erythrocytes.

The relationship between recurrent oral ulceration and coeliac disease has been discussed in Chapter 4. It is generally accepted that ROU is not generally associated with iron deficiency (as may occur in coeliac disease), although pre-existing ulceration may well be exacerbated by it. It is less easy to define the role of folate or $B_{12}$ deficiencies—some deficient patients with ROU show an immediate response to supplementation whilst, quite clearly, those with coeliac disease respond primarily to the gluten-free diet. This, however, results in the correction of the malabsorption process and the restoration of deficient levels. It is, therefore, difficult to tell precisely what part each component plays in the reversal of the ROU.

Further uncertainty is introduced by the fact that some patients with advanced haematological disturbances do not show oral changes. As an example, some patients with advanced megaloblastic anaemia, the result of folate deficiency, may have no oral complaints and the mucosa may appear to be within normal clinical limits. It would seem reasonable to revise the long held view that clear-cut oral signs and symptoms may be ascribed to specific blood deficiencies, and to adopt the alternative view point that a wide range of oral changes may arise from any of the conditions under discussion.

In the majority of the patients described treatment of the blood abnormality leads to a rapid resolution or improvement in the oral symptoms. In patients with long-standing latent iron deficiency, however, the response may be a slow one. Such refractory behaviour is well recognized; it is often more difficult to restore serum iron levels than haemoglobin levels by oral therapy.

## Selection of patients for examination

Patients with the described symptoms form a considerable proportion of those referred to the oral medicine clinic for investigation and, as a consequence, a full haematological examination must be carried out in all these cases. The accepted routine 'blood screen' (consisting of a haemoglobin estimation, a full blood cell count, and the examination of a stained blood film) is insufficient to demonstrate anaemias in their early or latent stages and it is important to extend the investigation further in selected patients.

The suggested screening procedure has been previously outlined, but it may be helpful to repeat this in the present context. It is suggested that the following groups be considered for investigation.

(1) Patients with persistent oral ulceration.

(2) Patients with oral lesions with an atypical history or unusually resistant to treatment.

(3) Patients complaining of a sore tongue, a generally sore mouth, or abnormal taste sensation, even though no mucosal changes can be seen.

(4) All patients with candidiasis.

(5) Patients showing abnormalities following the initial screening.

When an extended investigation is decided upon, a reasonable scheme of investigation is as follows.

(1) Full blood count, haemoglobin estimation, and film examination. This is the routine screen procedure, and from it evident anaemias are demonstrated by variations in red cell morphology and lowered haemoglobin values. Abnormalities of the white cells are also shown.

(2) Estimations of serum iron, total iron binding capacity, and saturation. As described above, these provide an important test for latent iron deficiency.

(3) Serum $B_{12}$, serum folate, and red cell folate estimations. The necessity for carrying out both folate estimations has been discussed in Chapter 2.

(4) As an additional test an erythrocyte sedimentation rate measurement is useful as a non-specific guide to underlying pathological processes such as chronic inflammatory conditions or neoplasia.

It is quite evident that further, more specialized tests may be necessary for a full diagnosis of some of the patients involved. A number of these have been discussed in Chapter 2.

It is quite essential that the presence of some haematological abnormality should be followed up with appropriate investigations according to the clinical circumstances. Simple replacement therapy for a deficiency is unacceptable unless attempts are made to track down the basic cause. It must be said, however, that even after carrying out full investigations, a small number of patients presenting with oral symptoms attributed to deficiencies remain a diagnostic puzzle in terms of the aetiology of the deficiency.

## Leukaemia

Leukaemia represents a malignant proliferation of white cells, replacing their normal development in the bone marrow. This process may affect any of the white cell strains, but the most usual forms are lymphocytic, monocytic, and myeloid, depending on whether lymphocytes, monocytes, or granulocytes are involved. Each of these forms of leukaemia present either in an acute or a chronic form. In general chronic leukaemias are more common in older patients, acute leukaemias in younger patients.

It is not uncommon for oral symptoms to be the first indication of the presence of leukaemia, particularly of one of the acute types. The symptoms are of a hyperplastic gingivitis resulting in fragile red spongy gingivae which bleed spontaneously or following slight injury. In more acute cases, the hyperplastic nature of the gingivitis may not be evident and the condition may show as free spontaneous haemorrhage from the gingival margins (Fig. 9.2). The gingivae are highly susceptible to infection and a secondary acute ulcerative gingivitis is common. This susceptibility to infection occurs also in more chronic cases when the oral symptoms may consist of recurrent attacks of acute ulcerative gingivitis. These may occur without any of the hyperplastic changes mentioned above and so may appear to be not of great significance. Recurrence of acute ulcerative gingivitis is, of course, not uncommon in the otherwise healthy patient and would seem to be associated with a defective immune mechanism. However, unexplained or repeated recurrence of acute ulcerative gingivitis should be treated with suspicion and blood examination arranged in order to eliminate the possibility of the blood dyscrasia.

In more advanced cases of leukaemia, oral ulceration is common. The ulcers, produced by the breakdown of the tissues overlying deposits of leukaemic cells,

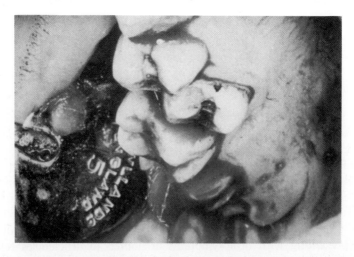

Fig. 9.2. Spontaneous haemorrhage from gingival margins in acute leukaemia.

may be large, painful, and difficult to treat. The maintenance of oral hygiene may be of great help in reducing the distressing oral symptoms in these patients. Covering agents and antibiotic mouthwashes are also helpful in easing the painful symptoms during the late stages of the disease.

## Leukopaenia

Leukopaenia represents a fall in the white cell content of the blood. This may be a spontaneously arising condition, but may occur also as a response to drug therapy. It may also occur as a transient stage in the development of leukaemia and other diseases affecting the blood marrow.

Although it is not a particularly common condition, the most usual clinical presentation of the kind is of agranulocytosis. This represents a reduction in the number of granulocytes formed in the marrow and circulating in the blood. The effect of this is to increase the susceptibility of the patient to infections of various kinds. In the case of the oral mucosa this may lead to widespread infection and ulceration of all parts of the mucosa. These changes may not be dissimilar to those occurring in leukaemia; the aetiological process is similar in that the protective function of the white cell component of the blood is reduced in the one case by the production of abnormal cells and in the other case by an inadequate production in terms of numbers.

There is currently considerable interest in the role of neutrophil function in periodontal disease. It is suggested that, whilst neutrophil activity may play a part in the destructive immune processes of chronic periodontal disease, deficiency in numbers, or in function may also result in periodontal lesions. It is certainly the case that in cyclic neutropenia (a condition affecting young patients in which neutrophil production is intermittently deficient) marked gingivitis and oral ulceration may occur during the neutropenic episodes.

## Platelet abnormalities

When the function or the number of the platelets in the circulating blood is reduced there is a tendency for spontaneous haemorrhage to occur within the tissues. This may well show initially in the form of petechial haemorrhages on the oral mucosa and these are, in fact, a well known sign of early idiopathic thrombocytopaenia. In leukaemias of various kinds platelet function is also greatly reduced, and haemorrhages of mucosa and skin may also be an early warning sign. It is advisable, therefore, to carry out a full blood screen, including a platelet count, on any patient with otherwise unexplained areas of haemorrhage in the oral mucosa. It should be remembered, however, that transient haemorrhages of this kind may occur on the soft palate in patients with a severe cold. Oral blood blisters (without any associated haematological defect) were described in Chapter 8. Similar blood filled blisters may also be produced in thrombocytopenia, particularly on the palate, although they may occur elsewhere on the oral mucosa.

## Nutrition

The integrity of the oral mucous membrane is maintained by a wide ranging complex of factors including those dependent on adequate nutrition. The significance of iron metabolism and associated factors has been previously discussed as has the relationship between gastro-intestinal disease and nutrition.

A wide range of conditions has been described in the past which depend on the absence or reduction of certain specific nutritional factors, particularly vitamins. With a few exceptions these specific conditions are now rarely seen under European conditions although this may certainly be far from the case in other situations. It should be remembered that a nutritional deficiency may occur in three ways: (1) as a result of reduced intake; (2) as a result of faulty absorption or metabolism; and (3) as a result of increased excretion. The relationship of iron deficiency anaemia to these three factors is a good and simple example.

Most patients seen in the oral medicine clinic with folate deficiency have this as a result of some form of malabsorbtion rather than poor intake. It should be remembered, however, that a high alcohol intake may result in low folate levels as may some drugs—phenytoin in particular. Only rarely, and then usually in strict vegans, is $B_{12}$ deficiency a result of poor dietary intake. On the whole, those patients who adopt unusual diets do so on a reasonably informed basis and are found to have satisfactory haematiological indices.

There has recently been considerable interest in the role of other nutritional elements in the integrity of the oral mucosa—the B complex of vitamins and a number of trace elements (particularly zinc) have been the subject of investigations. As yet, however, no major findings in these fields have been published.

It has previously been pointed out that in patients with nutritional deficiencies, secondary effects may follow. Predominant amongst these is the suppression of the normal immune response. This has been described in Chapter 3 in the case of nutritionally deficient children affected by cancrum oris. Nutritional deficiencies dependent on faulty diet are rarely simple ones and the patient suffering from any specific deficiency should be considered a candidate for a more complete investigation of nutritional standards.

## Scurvy

Scurvy (ascorbic acid deficiency) is now an uncommon disease in Europe, but is not by any means unknown, being the most commonly recognized condition associated with a single vitamin deficiency. Although the disease occurs more often in old and neglected patients, there are occasional cases of much younger individuals who adopt such a restricted form of diet that clinical signs of ascorbic acid deficiency appear.

The leading oral symptom is a hyperplastic gingivitis, the gingivae becoming swollen and friable, and purple-red in colour (Fig. 9.3). There is marked false

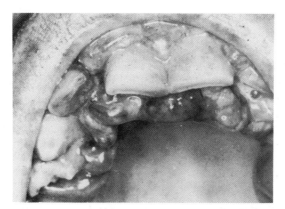

**Fig. 9.3.** Severe hyperplastic gingivitis in scurvy.

pocketing and this, together with general lack of tissue resistance, may lead to secondary infection. There may be, therefore, a superimposed acute gingivitis. Generalized symptoms include tiredness and malaise. Capillary fragility is a feature of this condition, and may lead to the appearance of spontaneous haemorrhage and widespread bruising, particularly around the joint areas.

Investigation of such a patient must include a blood examination in order to eliminate the possibility of a blood disease, since the most important differential diagnosis is from leukaemia. The main blood change in scurvy is the presence of a secondary anaemia. The positive diagnosis is by laboratory tests for leukocyte or plasma ascorbic acid levels, or by measuring the urinary output of ascorbic acid after a test dose. However, a satisfactory clinical diagnostic test is derived from the response of the symptoms to therapeutic doses of vitamin C, a response which occurs within a few days and which is accompanied by a dramatic reversal of all the symptoms.

The treatment of scurvy is a general medical and, sometimes, social problem. Although the administration of a high dose of ascorbic acid (1 g daily) for a few days may improve the condition of the patient remarkably, the further management must be a matter for the general medical practitioner. Not only the deficiency state itself, but also the conditions leading to its appearance must be corrected. It should also be remembered that although the symptoms presenting may be of ascorbic acid deficiency there is every likelihood that in fact multiple dietary defects exist.

## Endocrine changes

In general, changes in the oral mucosa dictated by endocrine abnormalities are not common. Perhaps the most frequently observed changes are those due to the endocrine disturbances found in normal life—at puberty, during pregnancy, and at the menopause. However, a few well established oral changes occur in some endocrine pathologies and these will be outlined below. It should be remembered that endocrine disorders are highly complex and often involve a number of systems because of the feedback mechanisms which control the endocrine system as a whole. It is, therefore, often difficult to determine the exact effect of a single endocrine abnormality on any structure.

## Pregnancy gingivitis

During pregnancy, the hormonal changes which occur may have the effect of exacerbating a previously existing chronic gingivitis which may have been previously symptom-free and unrecognized. The resulting gingivitis is essentially hyperplastic although there is minimal proliferation of fibroblasts. The most marked proliferation is of capillaries and this leads to the typically purple coloration of the gingival papillae (Fig. 9.4). These papillae tend to be fragile and may bleed at the least injury. Because of the presence of false pocketing and bleeding, stagnation and secondary infection may occur, and may lead to halitosis.

Occasionally, a single papilla may become considerably enlarged and present as an epulis-like growth. This is the so-called pregnancy tumour. In spite of its name this consists only of immature granulation tissue and histologically resembles the structure of the individual papillae affected by the more diffuse pregnancy gingivitis. The clinical characteristics and timing of the occurrence of these pregnancy lesions is sufficient to give a strong presumptive diagnosis. However, should any doubt occur as to the nature of the condition, full investigation is essential, including, if need be, biopsy excision of any doubtful over-growth. Pregnancy is, in itself, no contra-indication to such a biopsy, but it should be remembered that the lesion is likely to be extremely vascular and profuse bleeding is to be expected. On balance, it is often better, if a confident

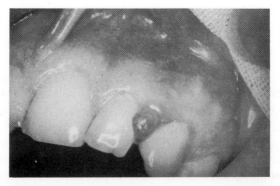

**Fig. 9.4.** Pregnancy gingivitis. The discoloured papilla is typical of the condition.

clinical diagnosis has been made, to avoid biopsy since the condition is likely to regress considerably, if not completely, after pregnancy.

Treatment during pregnancy should consist of the application of strict oral hygiene measures. This, in itself, is often sufficient to halt the progress of the gingivitis. However, oral hygiene measures alone are unlikely to lead to the complete resolution of a discrete epulis-like mass and eventual excision may be necessary.

## Puberty

During puberty, particularly in females, a similar series of changes to those in pregnancy may occur. These changes correspond to an increased secretion of the sex-determined hormones. Just as in pregnancy it is suggested that the gingival changes of puberty may be minimized if not completely eliminated by adequate oral hygiene measures. It need hardly be said that the oral changes of puberty are self limiting and self reversing.

## Menopause

During and immediately after the menopause oral symptoms are common. These chiefly take the form of soreness of the oral mucosa associated with variations of sensation and of taste. There may be in some instances apparent atrophic change in the oral mucosa itself although frequently no abnormality can be seen. Very often minor changes of this kind are accentuated by the patient who may show an exaggerated response. These menopausal changes are self limiting but are often the cause of great distress and may result in a great deal of investigation with little factual result. By and large, patients in this group with no clearly defined aetiological factor for their diffuse stomatitis are best treated by reassurance and minimal or placebo action. In the past a number of oral conditions have been directly ascribed to the changes of the menopause, in particular 'desquamative gingivitis', but it is now realized that there is no evidence for any hormonal involvement.

It should be pointed out, however, that there is always the possibility of the onset of some such conditions as lichen planus or an anaemia which may initially show in precisely the same way as the non-specific mucosal changes of menopause.

## Addison's disease

Addison's disease is the result of lack of function of the adrenal cortex, usually the result of an autoimmune disorder, but with other possible aetiologies. As a result of this destruction the feedback mechanism between the adrenals and the pituitary is disturbed and a wide ranging series of endocrine changes results. The oral change of significance in Addison's disease is melanotic pigmentation of the oral mucosa, which may include buccal mucosa, gingivae, and palate. It should be remembered that this form of pigmentation is by no means specific, but none the less the appearance of such signs in a patient known previously to be free of pigmentation should always be considered as of significance. The mechanism of melanin production in this way is not clearly known. Although a melanin-stimulating hormone is secreted by the pituitary gland it seems that the actual onset of pigmentation may be associated with variable ACTH levels. In more fully developed Addison's disease oral candidiasis may also occur. The association of oral melanotic pigmentation with candidiasis is evidently a further, more marked indication for endocrine studies.

Although Addison's disease is a well known cause of oral mucosal pigmentation it is, in fact, very unusual for the disease to be recognized in this way. There are many much more common causes of oral pigmentation. Some of these are considered below.

## Hypothyroidism

Excessive production of thyroid hormones does not have any direct effect on the oral mucosa. However, decreased production (hypothyroidism) is associated with an impaired immune mechanism. Oral candidiasis (as well as cutaneous candidiasis) may be the result. These mucosal changes, in association with swelling and oedema of the facial tissues are characteristic of acquired hypothyroidism (myxoedema). If gastric parietal cell autoantibody screening for pernicious anaemia is carried out, it is probable that antithyroid antibodies will also be estimated. These are often present in pernicious anaemia, and there is a relationship between pernicious anaemia and hypothyroidism—possibly in the family if not in the individual patient.

Congenital hypothyroidism (cretinism) includes enlargement of the tongue and irregular swelling of the gingivae among its manifestations.

## Diabetes mellitus

Diabetes mellitus has no specific oral signs or symptoms. However, possibly because of the general lack of resistance to infection of the diabetic patient, periodontal disease processes may become exaggerated and it is not uncommon to find that such a patient presents with advanced periodontal disease. The principles of treatment of this are simply those of periodontal treatment in the non-diabetic patient. However, in this and in all other treatment the dental surgeon must always remember that the diabetic patient is more than normally susceptible to infection. This is true whether or not the diabetes is fully controlled by medication. It is, therefore, necessary to carry out any form of surgical treatment under antibiotic cover and to recognize the possibility of infection in situations in which it would normally not be anticipated.

The patient with undiagnosed or inadequately treated diabetes may have a generalized stomatitis and, in particular, a sore tongue. This is probably at least partly due to dehydration and partly due to candidal infection. The possibility of latent diabetes should be considered amongst others in the patient with non-specific glossitis or a candidiasis; a family history of diabetes is of particular significance. Just as in the case of the sore tongue due to anaemia, a sore tongue resulting from diabetes may occur early in the disease process before substantial amounts of glucose are passed into the urine. In order to provide a sufficient screening procedure it may be necessary in these patients to carry out a fasting glucose estimation or glucose tolerance test rather than a simple urine screen.

# Oral reactions to drug therapy

## Antibiotics

It is well known that hypersensitivity reactions may occur during antibiotic therapy. These are commonly generalized reactions involving the whole metabolism. Their severity may vary from a mild and transient rash to the extremely severe reaction of angioneurotic oedema, in which oedema and swelling of the tissues of the head and neck may extend to the tongue and larynx and result in dangerous respiratory obstruction. Such a condition represents an extreme medical emergency and must be treated as such.

Hypersensitivity reactions may occur occasionally following the repeated use of tetracycline mouthwashes. The reaction may occur either early in the initial course or after many treatments. The reaction often takes the form of a localized angioneurotic oedema with swelling of the eyelids and the facial tissues in general. If this reaction occurs, it indicates that the patient has acquired a hypersensitivity to the tetracycline and must be warned that future use of the drug may be dangerous. The possibility of this type of reaction is yet another argument against the indiscriminate use of antibiotic therapy. Occasionally, a localized form of hypersensitivity reaction may be seen in the oral tissues following tetracycline mouthwash therapy. This is relatively limited in nature and leads to multiple vesicle formation. Again, the appearance of these symptoms should lead to the immediate cessation of the use of the antibiotic together with warnings as to further use.

Treatment of hypersensitivity reactions depends on the severity and acuteness of the symptoms. In a mild reaction it may only be necessary to discontinue the use of the drug and to observe the patient carefully. In somewhat more severe cases the use of antihistamines alone will be sufficient to suppress the symptoms. However, in a fully developed angioneurotic oedema immediate treatment with intravenous hydrocortisone, often combined with subcutaneous adrenalin may be necessary. Admission to hospital is essential in these grave cases, but initial treatment by 100 mg of hydrocortisone given intravenously may be life-saving.

Apart from these manifestations of hypersensitivity there are a number of minor reactions to antibiotics which may be localized in the mouth. Of these, perhaps the most common is the occurrence of black hairy tongue following treatment with either wide- or narrow-spectrum antibiotics. As the essential feature of black hairy tongue is elongation of the filiform papillae and this is associated only with secondary bacterial changes, it is difficult to understand why antibiotic therapy should induce this particular form of reaction. However, although there have been few controlled studies it would seem quite clear on clinical grounds that from time to time black hairy tongue does, in fact, follow antibiotic therapy. Following the cessation of the therapy the tongue may return to normal either quickly or, occasionally, very slowly. This form of reaction may occur after either systemic or local antibiotic therapy.

It might well be thought that the use of antibiotic mouthwashes to treat oral lesions would lead to symptoms arising from the widespread local overgrowth of resistant organisms and in particular, yeasts, but in fact, this is not often so. Precisely as general physicians have found it unnecessary to invorporate antifungal antibiotics with tetracycline in order to avoid gastro-intestinal infestations by yeasts, so it has been found unnecessary to combine antifungal antibiotics with wide-spectrum antibacterials in mouthwash therapy for oral ulceration. This does not mean, however, that such an overgrowth is not possible and it must always be remembered that any localized use of an antibiotic may leave behind resistant strains of organisms, even if no clinical overgrowth occurs. It is, therefore, worth reiterating that the use of antibiotic mouthwashes should be reserved for situations in which there are positive indications.

## Steroids

There are now many patients taking systemic steroids on a long-term basis and often this treatment results in susceptibility to infection and general loss of tissue resistance. In the mouth, this takes the form of acute atrophic candidiasis in most cases, although acute pseudomembranous candidiasis—thrush—may also occur. The mucosa is reddened and sore and there are atrophic changes in the epithelium which may result in localized erosions. Although the palate is most often affected the whole of the oral and pharyngeal mucosa may become involved as may be the mucosa of the larynx. Secondary infection by *Candida albicans* is almost invariably present and treatment of these patients by local antifungal therapy is often satisfactory in reducing symptoms. However, this can be no more than a temporary solution since these patients are, in general, destined to maintain steroid therapy for an indefinite period. It may, therefore, be necessary repeatedly to treat this candidal infection. In these persistent cases of candidal infection the use of miconazole may have considerable advantages over the more usual antifungal preparations.

Antibiotic-steroid mouthwashes are advocated for many conditions; in the higher concentrations these may replace systemic steroids. This form of therapy might be expected to be a prolific source of atrophic candidiasis, but in fact, this complication is relatively unusual. If it does occur long-term antifungal measures must accompany the antibiotic-steroid therapy. Miconazole gel is often very helpful in these circumstances.

As has previously been pointed out, patients on extremely high doses of steroids, possibly in conjunction with other immunosuppressive drugs (in particular those undergoing rejection following organ transplantation) may develop oral infections by organisms normally counted to be non-pathogenic. It has also been pointed out that such patients are much more than normally likely to develop neoplasms. Figures 9.5 and 9.6 show such a patient on extremely high dosages of immunosuppressive drugs who developed a crusted lesion of the lower lip which was found to contain a wide range of organisms, and also an adjoining leukoplakia of the labial mucosa. Both these lesions quite rapidly resolved on the cessation of the high dosage immunosuppressive therapy.

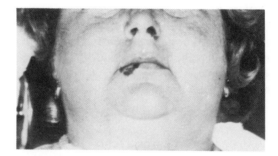

**Fig. 9.5.** Patient on intensive immunosuppressive therapy showing Cushingoid changes and an infective lesion of the lower lip.

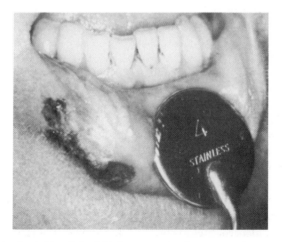

**Fig. 9.6.** Leukoplakia of the labial mucosa in the patient shown in Fig. 9.5.

## Sodium phenytoin (Epanutin)

Sodium phenytoin is the most commonly used drug for the treatment of epilepsy and, in approximately 50 per cent of the patients taking it, there is a marked chronic hyperplastic gingivitis. The hyperplastic reaction is typically papillary and the interdental papillae become swollen, sometimes grossly so (Fig. 9.7). This is essentially a fibroblastic reaction and the tissue is generally firm and much less haemorrhagic than in the case of pregnancy gingivitis. However, in view of the marked false pocketing which may occur, there are great difficulties in maintaining oral hygiene and secondary inflammatory changes are almost invariably present. The nature of the reaction to the drug is not clear since there is evidence that the hyperplasia represents an enormously exaggerated form of chronic gingivitis which does not occur if immaculate standards of oral hygiene are maintained. Although withdrawal of the drug is, in itself, sufficient to halt the progress of the condition, the difficulties of stabilization in an epileptic patient make it impractical to suggest its withdrawal as a treatment for the oral condition. Stringent oral hygiene and regular scaling preceded, if need be, by gingivectomy must be carried out. In severe cases, the gingivectomy should be performed under hospital conditions. Although the proliferation is essentially fibroblastic, the secondary infective processes often result in the production of very vascular tissue and blood loss may be considerable.

## Cyclosporin A

This is an immunosuppressive drug which is now widely used in all branches of transplant surgery. One of its side effects is the stimulation of a gingivitis very like that induced by phenytoin (Fig. 9.8). This also seems to be a fibroblastic reaction, but the density of the hyperplastic gingival tissue is much less than in

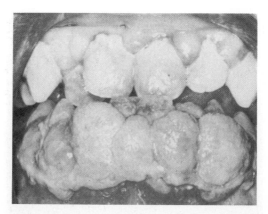

**Fig. 9.7.** Epanutin-induced gingivitis. The papillary nature of the hyperplasia is characteristic.

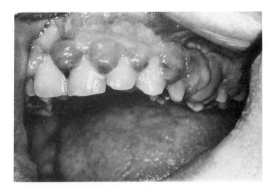

**Fig. 9.8.** Cyclosporin A—induced gingival hyperplasia.

phenytoin gingivitis. It is said that immaculate oral hygiene will prevent this gingivitis but this has proved almost impossible to attain in the authors experience. A similar reaction is now being reported as a response to nifedepine—an antihypertensive drug also often used in patients with renal failure.

## Aspirin burn

It should always be remembered that one of the most important common adverse reactions of the oral mucosa to drugs is that of trauma following contact with aspirin or an aspirin-containing tablet of some kind. Patients frequently use this form of therapy for the relief of toothache and the association of a carious tooth causing toothache and an acutely presenting white patch of the adjoining buccal mucosa should bring this possibility strongly to mind (Fig. 9.9). The

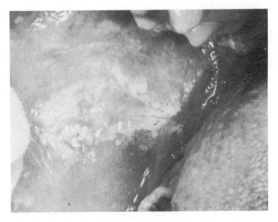

**Fig. 9.9.** Aspirin burn of the buccal mucosa.

appearance of the burned area may be quite spectacular and only the history will provide the diagnosis. The condition is, quite evidently, self-limiting and requires no treatment except for the toothache.

## Lichenoid reactions

It has been pointed out in Chapter 8 that lesions resembling lichen planus may be precipitated by a wide range of drugs. Similarly, lupoid and pemphigus, and pemphigoid-like reactions may occur in the same way and in response to a similarly wide range of medications. It may be very difficult to recognize, in any particular case, whether a drug reaction is involved. However, as previously pointed out, the drug-induced reactions behave much as do the non-induced diseases; withdrawal of the stimulating drug does not always mean that the reaction will subside. Management is the same in both situations. Recent observations of the apparent induction of lichenoid reactions by amalgam filling material are interesting, but at the present time it is very difficult to confirm such an association, even in the very few patients involved.

## Melanotic pigmentation of the oral mucosa

The significance of oral pigmentation in endocrine disturbances, and particularly in Addison's disease, has been discussed above. A number of drugs may also stimulate increased pigmentation—oral contraceptives, antimalarials, and tranquillizers among them. Increased melanin production may also occasionally be seen in association with the oral lesions of lichen planus and with leukoplakias; this is of no clinical significance. However, in many patients, particularly those with heavily pigmented skins, oral pigmentation is quite normal. This may be patchy or diffuse, but the gingivae are almost always involved, even when the skin pigmentation is minimal.

The recognition of a localized and symptomless melanotic patch on the oral mucosa may cause considerable problems. Such a lesion may be entirely benign and static, but the initial stages of malignant melanoma may have a similar appearance. The first step must be to eliminate the possibility of an amalgam tattoo; this may be difficult since a long-standing amalgam deposit may be diffuse and difficult to recognize on X-ray. If it is felt that the possibility of a malignant melanoma is a real one, wide excision biopsy should be carried out under conditions which allow for widespread surgery if found necessary during the biopsy procedure.

# 10

# Disorders of bone

Diseases of bone may be divided into three groups: congenital, metabolic, and inflammatory. Relatively few of any of these present with any frequency for investigation in the oral medicine clinic. Most bone disorders are diagnosed on a quite different basis. However, some genetically determined diseases have an impact on dentistry in general, often as part of generalized syndromes involving bone and epidermal appendages (including teeth). In these complex syndromes the predominant oral condition is variation in the size, number, morphology, and, sometimes, structure of the teeth. Cleido-cranial dysostosis is an example of one such syndrome in which the numbers of teeth are largely involved, osteogenisis imperfecta one in which the structure of the teeth may be affected. These factors have been discussed in Chapter 6.

The metabolic bone diseases most often appearing in the oral medicine clinic are fibrous dysplasia and Paget's disease. These are discussed below together with some other less often seen conditions. Inflammatory bone disease is only marginally within the field of oral medicine although osteomyelitis of the jaws is a classic example, now—in the antibiotic age—very rare. The relationship between hyperparathyroidism and the production of oral giant cell lesions has been mentioned in Chapter 7. The screening procedure which is likely to provide the first evidence of a metabolic bone abnormality is an estimation of serum calcium, phosphorus, and alkaline phosphatase, routinely reported in a multi channel blood analysis. These results will be further discussed below. Table 10.1 gives details of the levels in the conditions mentioned in this Chapter.

**Table 10.1.** Blood chemistry in disease of bone

|  | Ca | P | Alk. Phos. |
|---|---|---|---|
| Normal | 2.2–2.7 | 0.8–1.4 | Variable (see below) |
| Paget's disease | N | N | ++ |
| Monostotic fibrous dysplasia | N | N | N |
| Polyostotic fibrous dysplasia | + | N | + |
| Hyperparathyroidism | + | − | + |

Ca and P in mmol/l.
Alkaline phosphatase values. Normal levels for the age group should be determined from the specific laboratory. Usual adult values up to 375 IU.
N represents normal.
+ Represents a moderate rise; ++ represents a marked rise; − represents a moderate fall.

When interpreting the results for alkaline phosphatase it is important that the particular laboratory normals be used as a point of comparison; these vary more than most values. This variability is further complicated in younger patients by the wide individual variations which occur whilst active bone formation is occurring. In older patients it should also be remembered that there is an alternative source of alkaline phosphatase—from the liver. It is necessary to ask for an electrophoretic separation to be carried out if there is doubt on this score.

## Monostotic fibrous dysplasia

This is a condition which may arise in either male or female patients and is associated with very little other disturbance of bone or any other tissue. The lesions occur more often in the maxilla than in the mandible, and are reputed to remain within a single bone and not to cross suture lines. The essential change in this condition is the replacement of the normal bone architecture by a partially calcified fibrous mass with a histology suggesting an acceleration of the normal bone metabolism of osteoclasis and osteogenisis. This process is not associated with any generalized change; in particular, the accepted blood chemistry determinants (calcium, phosphorus, and alkaline phosphatase) remain unchanged. The degree of ossification of the lesion is widely variable, tissue removed from a lesion may vary from a very soft and haemorrhagic specimen to a relatively hard and well ossified tissue. In the case of well demarcated lesions of fibrous dysplasia, there is a great deal of discussion as to whether these represent neoplastic changes of a benign type. This question has been, to some extent, side-stepped by the adoption of the term 'fibro-osseous lesion' which refers to lesions of this kind.

The clinical presentation of fibrous dysplasia is of an otherwise symptomless swelling of the mandible or maxilla (Fig. 10.1). In the case of the maxilla, the swelling may encroach on the antral cavity (Fig. 10.2). There is virtually no

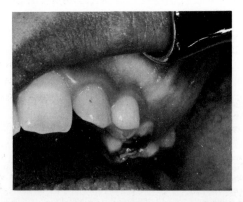

**Fig. 10.1.** Fibrous dysplasia. A monostotic lesion of the left maxilla.

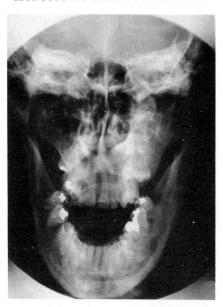

**Fig. 10.2.** Fibrous dysplasia. Radiograph of lesion in left maxilla showing thickening of bone and loss of antral cavity.

other complaint and all investigations of a biochemical nature prove to be unproductive. The only useful investigation is radiography, although the appearance of fibrous dysplasia may be very variable. In general, the basic pathological process of decalcification and recalcification is reflected in a mottled appearance of the bone on X-ray. This, however, depends on the stage and the rapidity of the process. In those patients in whom the process is slow and includes a significant element of calcification the mottling will be minimal and the expanded bone will have an almost normal appearance. If the decalcification of bone is predominant at the time of X-ray then the radiographic appearance will reflect this fact. The only absolute method of diagnosis is by biopsy; it has in the past been suggested that this is a dangerous procedure leading to onset of malignancy, but this is a suggestion which has never in any way been proved. So far as can be ascertained, biopsy in fibrous dysplasia is a perfectly safe procedure, the only difficulty being of haemorrhage in the more fibrous and highly vascular phases of the condition.

Treatment of monostotic fibrous dysplasia should be highly conservative. There is no medical treatment available and, since the lesions progress to a self-limiting static phase over a few years, it is almost always better to await events before carrying out surgery. The extent of this surgery should be determined entirely by cosmetic factors. Since neoplastic change is virtually unknown in this condition, there is no need to attempt to remove the whole of the lesion and, in fact, this is often an almost impossible task. It is generally accepted, therefore, that simple cosmetic contouring of the facial skeleton at the latest possible moment, when the lesion might have been expected to run its full course, is the proper treatment.

## Polyostotic fibrous dysplasia

This condition is very much less common than is monostotic fibrous dysplasia. It is virtually always associated with widespread changes throughout the skeleton and in other systems of the body which are collectively known as Albright's syndrome. In this condition areas of bone throughout the body are replaced by fibrous tissue with widely differing amounts of new ossification included within them. This often leads to multiple fractures and to gross distortion of the skeleton. In the full Albright's syndrome, the bony lesions are associated with patchy melanotic skin pigmentation and, in the case of females, sexual precocity. This condition is a far more active one than that of monostotic fibrous dysplasia and this is reflected in the blood chemistry changes. In polyostotic fibrous dysplasia the serum alkaline phosphatase is often greatly elevated, as in the serum calcium. This, of course, is in contrast to a situation in monostotic fibrous dysplasia when no such changes can be shown. Although polyostotic fibrous dysplasia may initially show as a condition of the jaws, it will rapidly become evident in other bones of the skeleton. The contrast with monostotic fibrous dysplasia is evident, polyostotic fibrous dysplasia is a generalized condition which also affects the jaw bones whereas monostotic fibrous dysplasia is essentially a single lesion affecting the jaws without any other skeletal or other generalized abnormality.

## Paget's disease

Paget's disease is a widespread condition of old age, being described as present in a large proportion of older patients coming to autopsy. Such changes are quite asymptomatic and may be distinguished from the obvious changes which occur in the clinically significant form of the disease. Rather as in fibrous dysplasia, Paget's disease represents an imbalance of the osteogenic and osteolytic processes occurring in normal bone formation. The final result is bone growth. This often occurs initially in the skull and facial bones, and it is, thus, often recognized in the oral medicine clinic. The classic complaint of a patient is of hats of too small a size, but an equally common complaint is of dentures which are becoming 'too small'. The bone growth in Paget's disease particularly affects the vault of the skull and maxilla although the mandible may also be involved (Fig. 10.3). The expansion of the bone of the base of the skull leads to closure of the foramina and resultant neurological changes such as deafness. Nerve compression may also lead to neuralgia-like symptoms in the trigeminal nerve; this

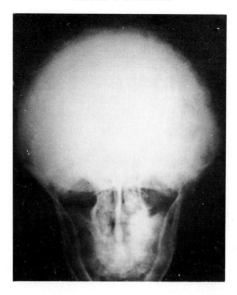

**Fig. 10.3.** Paget's disease. Radiograph showing expansion of the skull and 'cotton wool' appearance of the bone.

should always be considered as a possible diagnosis in older patients, particularly when there are other symptoms such as deafness. Apart from neuralgia-like symptoms the patient may complain of pain within the bone itself. This is a common symptom known as bone pain. Radiographically the bone is reputed to show the appearance of cotton wool, and, in fact, this is quite often the case. Since cementum is essentially bone this is also affected by the changes of Paget's diseases and hypercementosis is a common finding in these patients. It has been suggested that malignant change in the affected bone is to be expected although the incidence must be relatively low. Cardiac failure because of high output into the expanded blood spaces of the bone is a well recorded complication. It is not so certain that it occurs with any degree of regularity.

Diagnosis of Paget's disease is in the first instance clinical, confirmed by X-ray and by blood chemistry, the characteristic finding being a greatly increased serum alkaline phosphatase level. Other biochemical tests are available, in particular, the urinary levels of hydroxyproline, a measure of the change in collagen metabolism associated with the abnormalities of calcification.

In the past there has been no satisfactory treatment for Paget's disease, but recently two forms of therapy have been introduced. In the first of these fluoride supplements have been used to fix the calcium in the bone. This is shown to be effective by radiographic and clinical criteria, in particular by the reduction of bone pain and of peripheral nerve pain. More recently, a hormone, calcitonin, has been used. This thyroid-derived hormone acts directly on the bone resorption process and is often very helpful in arresting the changes of Paget's disease. The calcitonin used is derived from pigs, salmon, or as a result of synthetic processes. More recently still, the use of disodium etidronate has been shown to be effective in the reduction of symptoms: it inhibits bone metabolism as do other members of its chemical group.

Apart from the diagnosis and treatment of Paget's disease the most important dental consideration is that extraction of teeth in these patients may be difficult because of the hypercementosis mentioned above. Surgery may be followed by profuse haemorrhage because of the irregular nature of the blood supply to the new bone. Patients with Paget's disease are also notoriously susceptible to infection following any form of intervention and any extraction or oral surgery should be covered by antibiotic treatment. These problems of haemorrhage and infection lead to great caution in taking biopsies in Paget's disease, since this is a condition which can be diagnosed by non-interventive means the question of confirmatory biopsy should be approached with great reserve.

## Other bone diseases

As has been pointed out above, a very few other diseases of bone are of significance, for one reason or another, in the oral medicine clinic.

Rickets is a condition which occurs in children as a result of deficient calcification of the bones and (unusually) the teeth. It is essentially due to lack of vitamin D, either because of nutritional deficiency, malabsorption, or impaired metabolic processes. In this condition the bones are poorly formed and, as a result, badly shaped. It is said that the teeth are not affected in this condition, but in fact, some patients with a history of rickets have signs of hypoplasia of the teeth (Fig. 10.4). Osteomalacia is the adult equivalent of this condition which may occur in pregnancy, in malabsorption (such as in coeliac disease) or in renal disease. Its appearance in the oral medicine clinic must be rare.

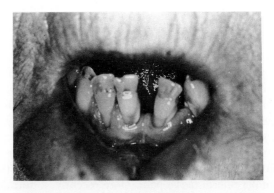

**Fig. 10.4.** Hypoplasia in a patient with a history of rickets.

Osteoporosis is a condition in which the skeletal bone structure undergoes degradation both of bone matrix and calcium; that is, a reduction in bone mass. There is skeletal rarefaction and crush fractures often occur. Apart from age changes the most likely cause of osteoporosis encountered in the oral medicine clinic is as a result of systemic steroid therapy. There are no fixed criteria as to the dosages to be used to avoid this effect. It is clear that it is necessary to use as low a dosage as possible in any given situation. The use of high concentration local steroid therapy rather than systemic therapy for oral lesions is the result of the consideration of this kind of side effect.

# 11

# Facial pain: neurological disturbances and temporomandibular joint

PATIENTS with problems of facial pain are seen daily in the dental practitioners' surgery. They are, in general, diagnosed and treated with relative ease, but there remain some few patients in whom the origin of the pain symptoms remains obscure and who are referred to specialist clinics of various kinds for diagnosis. Apart from these patients with symptoms of pain a very few patients present with other neurological disturbances around the mouth and face. The diagnosis of these may be a complex matter and outside the province of dentistry, but it is important that the dental practitioner is able to recognize significant changes and initiate further action.

The majority of patients with orofacial pain are suffering from some easily detectable pathological process in the teeth, their supporting tissues, or associated structures. Whatever the nature of the pain and however unusual its presentation, the first step must be the elimination of the possibility of its origin in this way; the many possible local origins of pain such as abnormal pulps, buried teeth and roots, periapical lesions, and so on, must be, as far as possible, eliminated by clinical and radiographic examination. This may not be simple; facial pain of dental origin may be difficult to recognize and referred or projected pain may make localization difficult. Vitality testing of the teeth and the use of diagnostic local anaesthetic injections may be useful aids. In particular, the pain associated with temporomandibular joint dysfunction may be very puzzling and may simulate a number of other conditions.

The matter may be further complicated by the difficulties which may be met by the patient in describing his symptoms accurately. On the whole, however, the terms used by patients to describe pain—'stabbing', 'throbbing', 'dull ache', and so on—are fairly constant and related to the aetiology of the pain.

The reaction of the patient to pain depends on two factors independent of the strength of the pain stimulus. These are the pain threshold for the patient (the degree of stimulation necessary for the patient to perceive pain) and the individual sensitivity to the pain when perceived. These two factors vary greatly from patient to patient and, in the case of the individual sensitivity, may vary from time to time in the same patient, depending on general health and other transient factors.

## Projected and referred pain

If a pain pathway is subjected to stimulation at some point along its course it is possible for pain to be felt in the peripheral distribution of the nerve. This is projected pain. An example of this is the facial pain which may occur in patients with intracranial neoplasms. The lesion in this case is causing pressure on a central part of the trigeminal pathway, whilst the pain symptoms are felt in the peripheral distribution of the nerve.

This must be distinguished from referred pain in which the pain is felt in an area distant from that in which the causative pathology is located. In the dental field perhaps the most common example of referred pain is that in which pain is felt, for instance, in the maxillary teeth due to a lesion in a mandibular tooth. In this case the pain impulses originate in the diseased tooth and if the pathway from this tooth is blocked—say by a local anaesthetic—then the referred pain will cease. Thus, in our dental example, anaesthesia of the mandibular branch of the trigeminal nerve will relieve the pain felt in the area supplied by the maxillary branch. This gives the basis of the most useful test for the investigation of suspected referred pain.

The mechanism of projected pain is fairly clear, but that of referred pain is by no means so obvious. There is still considerable discussion among investigators as to the true site at which the 'cross-over' occurs leading to the sensory misinterpretation of the impulses transmitted from the diseased site.

## The nerve supply to the face

The sensory nerve supply to the face and oral tissues is shared between a number of nerves: the trigeminal nerve, the glossopharyngeal nerve, and the branches of the cervical plexus. However, the great majority of pain symptoms in the face are felt in the area covered by the trigeminal nerve. There is usually a clear-cut distinction between the zones supplied by the various terminal branches of the trigeminal nerve, with very little overlap, but there is often considerable overlap between the trigeminal and cutaneous branches of the cervical plexus where these are adjacent. Apart from areas of overlap there is also a complex series of interconnections between the trigeminal, facial, and glossopharyngeal nerves, and the nerves arising from the autonomic nervous system, in particular the sympathetic fibres associated with blood vessels severing the area. These sympathetic fibres may play some part in the transmission of deep impulses; it has been shown that such pain can be produced by the stimulation of the superior cervical sympathetic ganglion. Similarly, it has been shown that pain impulses from the deeper facial structures may be transmitted by the proprioceptive fibres of the facial nerve. However, there is little agreement among the various workers who have investigated these secondary pain pathways as to their clinical significance.

The great majority of patients complaining of pain in and about the face are suffering from some form of toothache. However, there are many other possible causes of such pain. The structures related to the mouth which might give rise to pain symptoms are so complex and the innervation of these structures so interrelated that errors in diagnosis are easily made. The main sensory nerve of the area, the trigeminal nerve, eventually divides into a large number of small terminal branches supplying the skin of a large part of the face and scalp as well as the majority of the oral tissues and many deeper structures. Although the superficial limits of the trigeminal nerve can be accurately determined with little overlap the deeper limits are much less well defined and understood. It is often difficult to determine the point at which a facial pain becomes a headache and there may be consequent difficulties in communication between the patient and the investigator.

It is convenient to consider facial pain as being of four types.

1. Pain resulting from recognizable pathological change in the oral, facial, and closely associated structures. Examples are toothache, the pain of maxillary sinusitis, and pain associated with temporo-mandibular joint disturbance.

2. Pain of unknown origin felt in the face. This includes trigeminal neuralgia and the so-called 'atypical facial pain'—a mysterious, but very troublesome condition which will be discussed later.

3. Pain projected to the face due to pathological change affecting the pain pathways central to the facial structures. Secondary neuralgia caused by pressure of a neoplasm on the trigeminal nerve is a rare example of this group. Migraine (not strictly a facial pain) could also be classified in this way because of its aetiology—changes in the intracranial blood vessels.

4. Pain referred to the face from distant areas of the body. An example of this are in angina pectoris, in which pain may be felt over the left mandible.

It should be stressed that there is no characteristic pattern of pain associated with any of these four groups. They form only an arbitrary series of headings under which facial pain may be considered when presenting for diagnosis and, in the subsequent discussions there will be seen to be a number of factors which must be regarded as common between the groups.

## The investigation of facial pain

The objective investigation of pain is extremely difficult. Pain in itself produces no detectable signs and the basis of the whole investigation is the description given by the patient. This may be so coloured by a variety of personal factors as to be confusing and misleading. All pain causes a mental reaction and it is often impossible to differentiate between 'normal' and 'abnormal' reactions because of a complete lack of any means of objective measurement of the intensity of the pain. The function of pain in giving an early warning of pathological changes is notoriously erratic, and the excessive degree of pain associated with many conditions seems quite disproportionate if regarded as a warning sign. Many of the clinical conditions seen by the dental surgeon seem to fall into this category in which minor (or even quite unrecognizable) pathological changes elicit a quite disproportionate response leading to great distress in the individual concerned.

The investigation of a patient complaining of facial pain is carried out in two stages. In the first stage an assessment of the pain is made relative to the dental apparatus (teeth and supporting tissues) and the closely related structures, such as the maxillary antra and the temporo-mandibular joints, this corresponding to the first group defined above. If, after full examination, no abnormalities are found in these areas the investigator is justified in considering other possibilities. If the essential first step of eliminating dental causes of the pain is omitted confusion is bound to follow in many cases.

However, when all these localized causes of pain are eliminated by careful investigation, there remain a number of conditions of less evident origin for consideration in a possible diagnosis. These include primary neuralgias, secondary neuralgias, migraine, and cluster headache. In addition, referred pain and the so-called atypical facial pain must be considered.

## Neuralgia

The definition of neuralgia depends on a number of rather diffuse clinical factors. It is perhaps best thought of as a pain localized to an area supplied by a single sensory nerve, paroxysmal and severe in nature. A number of conditions give rise to pain of this nature in and about the face. Many classifications are possible but the most important distinction is between primary and secondary neuralgias. Primary neuralgia implies a condition in which the pain is present without any known initiating pathology. This does not mean to imply that pathological change does not occur, simply that none has as yet been detected. The most common primary neuralgia is that affecting the trigeminal nerve in one or more of its sensory branches, but a very similar although less common condition affects the glossopharyngeal nerve. In secondary neuralgias precisely similar pain may occur, but pathological changes affecting the course of the involved nerve can be recognized. This pathological change may affect any part of the nerve, from the periphery to the central sensory nuclei, the pain being projected to the peripheral distribution. Thus, in the case of the trigeminal nerve secondary facial neuralgias may occur as the result of neoplasia, trauma or inflammatory change affecting the extra- or intra-cranial course of the nerve.

It should be pointed out that various clinical conditions occasionally arise in which pain similar to that described may occur, but which is not usually considered to be a neuralgia. For instance, pain of temporo-mandibular joint origin may occasionally simulate that of a neuralgia. Some authorities would consider this to be a secondary trigeminal neuralgia, but it is perhaps better to reserve this term for those conditions in which an extraneous pathological process impinges upon the sensory pathways of the trigeminal nerve.

## Trigeminal neuralgia

Trigeminal neuralgia is a disease of such characteristic intensity that it has been known and described over a long period of history. In spite of this, the true nature of this extremely painful condition is not known. Although pain is normally considered to be a symptom rather than a disease, the lack of other associated symptoms and of basic knowledge as to aetiology makes it at present necessary to consider the pain as being the disease itself. Numerous investigations into the nature of the disease have been carried out in recent years but, probably due to the extreme difficulty in obtaining material for histological or ultrastructural investigation, these have as yet proved fruitless.

Trigeminal neuralgia is a disease of later life; most patients are female (F:M = 2:1). The pain distribution is almost always (but not invariably) unilateral and, at least in the first instance, confined to one division of the trigeminal distribution, although later there may be a spread to two or even three divisions. The pain is of great intensity, is described as stabbing in nature, and lasts for only a few seconds. This transient attack may be repeated in a matter of minutes or of hours. There may be no apparent precipitating factor, but many patients have a 'trigger zone', a point on the face (often outside the area of pain distribution) where the lightest contact may initiate an attack. Between the severe pains there may be a dull background ache in the area or no pain at all. These attacks rarely occur at night. The so-called 'frozen face' is an attempt by the patient to defer attacks by limiting movement, and hence avoiding the triggering of the pain. Spontaneous remission for a matter of weeks or months may occur, but this is very rarely permanent.

In the very early stages of the disease there may be a period in which the pain is non-characteristic. It is at this time that the patients are most often seen by the dental surgeon for the investigation of what may be highly perplexing pain symptoms. It is also at this stage of the disease that the patient may ascribe the pain to toothache in what may appear to be an entirely sound tooth. Following the extraction of this tooth the pain may then be described as coming from a nearby tooth and extraction of this may also be demanded by the patient. It is not rare to find patients made virtually unilaterally edentulous in this way. Many eventual difficulties can be avoided if it is recognised that any patient presenting with this pattern of symptoms should be considered as worthy of full investigation before further extractions are considered.

Conversely, the situation may arise in which pain closely resembling trigeminal neuralgia may arise from some quite ordinary dental pathology such as a retained root or an unerupted tooth or even from unsatisfactory dentures. If there is any abnormality of this kind present it should be treated as the essential first step in the management of the patient.

As has been pointed out, trigeminal neuralgia occurs in later life. The appearances of symptoms resembling trigeminal neuralgia in a patient under the age of 40 years should be treated with deep suspicion as a possible first warning of the onset of multiple sclerosis. In this condition the neuralgia-like pains are often the first symptoms. It is suggested that patients of this age group and with these symptoms should be subjected to a careful neurological assessment.

Medical treatment of trigeminal neuralgia depends largely on the use of the drug carbamazepine (Tegretol). This is highly specific treatment for neuralgia and is not otherwise an analgesic. Because of this specificity, response to its use may be accepted as a diagnostic indicator. Other anticonvulsant drugs are occasionally used, but carbamazepine remains by far the most important therapeutic agent in trigeminal neuralgia. The initial dosage of carbamazepine is kept low (perhaps 200 mg in the morning, 100 mg at night) and gradually increased until control is attained. A maximum dose of 800–1000 mg daily should not be exceeded. Minor side effects vary, some patients become drowsy whilst some report gastro-intestinal reactions. In a very few patients, however, the much more serious side effect of agranulocytosis may occur. The white cell count may be monitored by repeated blood tests, but it would appear that when this reaction occurs, it may do so suddenly and without prior warning.

A further serious side effect of carbamazepine which occurs in a few patients is the occurence of a severe rash; just as in the case of agranulocytosis this may occur quite unexpectedly after a long period of treatment with the drug. If medical therapy fails or is contra-indicated by side effects surgical intervention is necessary.

The surgical approach varies in complexity from relatively simple peripheral procedures to high frequency selective coagulation of the nerve fibres in the region of the ganglion. The formerly commonly used peripheral alcohol injection of the affected nerve has now been largely superceded by cryotherapy. The more major procedures are at the present time greatly improved in efficacy, safety, and freedom from severe side effects compared to the situation of only a few years ago.

## Glossopharyngeal neuralgia

Glossopharyngeal neuralgia is a similar condition to that of the trigeminal nerve, but much less common. The pain is of the same nature as that in trigeminal neuralgia and is unilaterally felt in the oropharynx, sometimes with pain referred to the ear. Although the main component of the pain is described (as in trigeminal neuralgia) as stabbing or shooting in nature, there may also be an appreciable residual ache which may last for some time after the paroxysmal attack. The pain is often precipitated by swallowing, and there may be a trigger zone. If such a zone is present, relief of the pain may be achieved by spraying the area with local anaesthetic solution. Treatment with carbamazepine is successful in the majority of cases and confirms the diagnosis.

## Secondary neuralgias

Facial pain and headache may be symptoms of a wide variety of intra- and extracranial lesions such as neoplasms or vascular abnormalities. Facial pain may arise by involvement either by pressure or by malignant infiltration of the trigeminal ganglion of the peripheral branches of the nerve. Primary and secondary intracranial neoplasms, nasopharyngeal tumours, aneurysms, and cerebral epidermoid cysts are among the most commonly recorded of lesions of this type causing facial pain. Similarly, the lesions left following brain damage of any kind may be the source of facial pain. In cases of this type facial pain is only rarely the only symptom, although it may well be the earliest. Involvement of other cranial nerves and the presence of unusual symptoms such as anaesthesia or paraesthesia should at once raise the suspicion of the possibility of a lesion of this kind.

Another cause of facial pain arising from compression of the trigeminal nerve is in Paget's disease affecting the base of the skull. Some degree of closure of the foramina is common and the consequent stricture of the nerve may lead to a variety of pain symptoms. A further cause of trauma to the trigeminal nerve in Paget's disease may be the deformation of the skull as a whole causing compression of the sensory root of the nerve where it crosses the petrous bone. The pain often simulates trigeminal neuralgia, although aching pains may be sometimes reported. A similar compression of the auditory nerve may give rise to deafness and changes in the cervical spine may also lead to pain in the area served by the sensory nerves of the cervical plexus.

A more peripheral form of nerve compression leading to pain may be seen in post-traumatic fibrosis around the infra-orbital foramen following fracture of the anterior wall of the maxillary sinus with involvement of the foramen. This line of fracture is common in injuries to the maxillo-zygomatic complex. Very occasionally, a similar fibrosis may follow infection in the area, say following a periapical infection on a maxillary canine tooth. The pain produced is usually paroxysmal, but not of the severe nature of trigeminal neuralgia. It is certainly wise, if trigeminal neuralgia confined to the infra-orbital nerve is suspected, to consider the possibility of such a fibrous stricture.

## Post-herpetic neuralgia

Pain may occur before, during or after an attack of facial *herpes zoster*. The first warning of such an attack may be a severe burning pain occurring in the area of the eruption, up to 2 days before the vesicles appear. This pain in *herpes zoster* is much more severe than that in *herpes simplex* in which the symptoms are largely due to the secondary infection of ruptured vesicles (Chapter 3). Following an attack of *herpes zoster* scarring of the involved nerve tissue may occur, resulting in a neuralgia-like pain which may be long lasting, intense, and very resistant to treatment. Older patients with *herpes zoster* are particularly liable to contract a post-herpetic neuralgia of this kind which may be associated with symptoms of hyperaesthesia or paraesthesia.

Treatment of established post-herpetic neuralgia may be very difficult. It does not respond to carbamazepine and analgesics may also be ineffective. Steroids have been advocated but there is no confirmation of their value. In some cases it may be necessary to resort to the use of chlorpromazine or equivalent drugs. The only form of treatment counted as effective is preventive. All patients with *herpes zoster* should be vigorously treated with antiviral agents. Currently, local and systemic Acylovir are used in combination. It seems established that successful treatment of the primary lesion considerably reduces the risk of the subsequent development of neuralgia.

## Migraine

Patients with migraine and migrainous neuralgia (cluster headache) are occasionally seen in the oral medicine clinic for differential diagnosis. Although migraine is essentially a headache, it may also have a facial component, usually over the maxilla, and this may cause confusion. In most cases, however, the patients complain of an intense headache, intermittent in nature, and usually (but not invariably) unilateral. Each attack of headache may last for a number of days, and is often associated with various symptoms such as nausea and visual disturbance. The systemic upset is considerable and many patients need to take bed rest until the attack subsides. The history of the condition with its associated disturbances often gives a strong clue to its nature, as does the frequent presence of a family history. Some patients recognize a precipitating factor, sometimes of stress, sometimes of a foodstuff. The onset of migraine may occur at any age, but is most common during adolescence.

The cause of migraine is thought to be instability of the cranial arteries which first contract and then dilate during the attack. Treatment involves some form of stabilization of this vascular abnormality. Unitl recently, the most widely used group of drugs have been derived from ergot and have needed to be applied with great caution in view of their powerful effects. Another drug, methysergide, also relieves attacks in a significant proportion of cases, but may induce side effects including the very serious one of retroperitoneal fibrosis. More recently however, clonidine, a drug used in larger doses for the treatment of hypertension and which has a vascular stabilizing effect, has been used in dosages of 50–100 µg daily for the control of migranous attacks. Although this drug has considerable side effects when used in large dosages for the control of hypertension, these relatively small amounts used in the treatment of migraine do not seem to involve such problems.

## Acute migrainous neuralgia

This condition, closely associated with migraine, may also appear as a diagnostic problem in the oral medicine clinic or in the dental surgery. The aetiology of this condition is as in migraine, but the patients are much more commonly young males. During an attack there are repeated episodes of pain, usually of about 30 minutes in duration, with a variable interval between. The pain is usually located behind the eye, with extensions to the maxilla and to the temporal area. There may be weeping of the eye on the affected side and congestion of the nasal mucous membranes. The attacks often occur at night. The characteristic grouping of attacks of pain over a short period followed by long periods of remission leads to the alternative name 'cluster headache'. Often the pain responds to antimigraine drugs in the same way as migraine itself.

Pain of a migrainous type may also be associated with the strange condition known as the Melkersson–Rosenthal syndrome. In this syndrome swelling of the lips is associated with transient and sometimes repeated attacks of seventh nerve weakness and with temporal headaches. Some patients have a deeply grooved tongue and granulomatous lesions of the buccal mucosa of a kind similar to those in Crohn's disease are described in some subjects. The aetiology of this extraordinary syndrome is not known, but the temporal headaches seem to respond to treatment with clonidine in the same way as does migraine. This is in accord with the suggestion that the neurological symptoms in this condition are due to vascular instability of some kind.

## Referred pain

Although pain is frequently referred between differing branches of the trigeminal nerve, pain referred to the face from a distant area is relatively uncommon. The only referred pain of this type seen with any degree of frequency is that in coronary insufficiency when pain may radiate over the left side of the mandible. This is usually in company with the other well recognized symptoms of the condition, but a very few instances have been reported in which pain over the left mandible has been the first complaint of a patient who subsequently developed symptoms of myocardial infarction.

## Atypical facial pain

Atypical facial pain is the term used to describe the condition presented by a few patients in which the distribution of the pain is unanatomical or in which the nature of the pain may be quite inexplicable and apparently unrelated to any pathological process. Pain is often bilateral and widespread, and the patient may complain of pain radiating over a considerable part of the body, but emanating principally from some area in the face. The pain is usually described as being a severe continuous dull ache, but this is not invariably so.

Some of these patients have evident psychiatric problems, but by no means all. The problem is often an intractable one, both in diagnosis and in treatment. In fact, treatment of the pain itself may be well nigh impossible, all attempts at interference making the situation worse.

Many variants of atypical facial pain have been recognized. In his textbook on orofacial pain Mumford describes three types. In the first group he places the incompletely understood conditions which remain obscure because of failure in diagnostic techniques. In the second group he places patients in whom augmentation occurs, those more aware than usual of their face and mouth. He comments that although these patients may seem peculiar and irritating they should still be regarded as within the normal range. These patients may pass from one clinician to another in an attempt to be given a satisfactory diagnostic label for their problem. Into his third group Mumford places patients whose symptoms are not explicable on physiological or anatomical grounds, but are more easily considered as having a psychiatric basis. These include the patients mentioned above with widespread and often bizarre complaints of pain.

Treatments of these patients may be protracted and often lies with the psychiatrist. However, as has been pointed out, some patients with quite inexplicable pain may simply be suffering from undiagnosed pathology. Mumford also makes the important point that patients with established psychiatric disturbance may also suffer from entirely routine conditions which cause facial or oral pain.

## Psychogenic factors

As has been pointed out above, a considerable proportion of patients presenting with anatomically inexplicable pain symptoms are eventually diagnosed as having psychiatric problems. It has also been suggested that a high proportion of patients with facial pain more explicable on anatomical grounds and, in particular, temporomandibular joint muscle dysfunction symptoms, may in fact be expressing a depressive background. This may be relieved (as will the symptoms of facial pain) by antidepressant treatment, prothiaden (dothiepin) being the recommended drug. In a few patients minimal problems of sore areas of the oral mucosa, often demonstrably habit based, may become grossly inflated in their impact and obsessional psychopathy may follow.

Many other oral conditions have been suggested as having a psychiatric basis, not surprisingly so since the face and mouth are well known as having potent psychological implications. None the less, although many conditions (such as recurrent oral ulceration and lichen planus) have been put forward as having a strong psychological overlay this has been very difficult to prove. The problem of the sore and burning mouth has been discussed in Chapter 5. At least some patients with this problem are never shown to have a physical basis for their complaint.

It is quite clear that some more generalized psychiatric problems may be directly reflected in the mouth. The dental erosion associated with bullimia has been mentioned in Chapter 6 as has the occasional occurrence of gross artefactual damage to the oral soft tissues by exaggerated cheek or lip chewing. However, probably the most common reflection of generalized psychiatric problems met in the oral medicine clinic is that of the dry mouth resulting from long-term treatment for psychiatric disease. The difficulties of dealing with this situation have been discussed in Chapter 5.

## Temporal headache

Two conditions which may very occasionally be seen by the dental surgeon, although not strictly facial pains, are temporal arteritis and the auriculo-temporal syndrome (Frey's syndrome).

In temporal arteritis the pain is localized to the temporal and frontal regions, the area supplied by the superficial temporal artery. The pain is described as a severe ache, but paroxysmal pain is occasionally also described. In between attacks of pain the affected area may be very tender to the touch. The patients, most of whom are in older age groups, also complain of general malaise, and diffuse muscular and joint pains. There may also be degeneration of vision. The essential pathology of this condition is a generalized inflammatory lesion of the arteries which shows itself early in the superficial temporal artery.

It is important that this condition is recognized and treated—it may otherwise progress to cause irreversible damage to the optic nerve. Diagnosis is often made on clinical grounds, largely dependant on the evidently enlarged and painful artery. The ESR is raised, often greatly so. Absolute diagnosis depends on the finding of a typical giant cell inflammatory process in a biopsy of the affected vessel but, clearly, this is not an everyday procedure. Treatment is with high doses of systemic steroids, of the order of 50 mg of prednisolone daily, until symptoms disappear.

The auriculo-temporal syndrome is caused by irritation of the auriculo temporal nerve as it passes through the substance of the parotid gland in the early part of its course. The irritation may be the result of a wide range of pathological processes within the gland, both inflammatory and neoplastic or by surgical trauma. The symptoms are of pain in the distribution of the auriculo temporal nerve, usually described as burning in nature, and associated with excessive sweating and erythema in the area during eating. Between attacks of pain there may be anaesthesia or paraesthesia of the skin in the affected area. It is evident that, the condition being recognized, confirmation of the diagnosis and subsequent treatment depends on a full investigation of the involved parotid gland.

# Neurological disturbances

## Facial nerve weakness

Paralysis or weakness of the seventh cranial nerve (facial nerve) may result from either of two types of neurological lesion.

1. *Upper motor neurone lesions* are central lesions which affect the nerve fibres responsible for movement in the lower facial muscles. In such a lesion the muscles over the maxilla and mandible are weakened, but those of the forehead are unaffected. This is because the motor nerves to the muscles of the forehead are regulated by both cerebral hemispheres, with free crossing of nerve fibres. In contrast the muscles of the lower part of the face are innervated only by nerves controlled in the opposite hemisphere. In these lesions the emotional movements of the whole of the face may be unaffected since these are initiated by upper motor neurones separated from the pyramidal motor fibres which serve the facial muscles for normal function.

2. *Lower motor neurone lesions* affect both the upper and lower facial muscles, the forehead, and face are equally affected. In these cases the pathological process is peripheral to the facial motor nucleus in the pons. The most common cause of such a paralysis is Bell's palsy (discussed below) but a wide variety of lesions affecting the facial nerve below the base of the skull (including some in the parotid gland) may be responsible for symptoms. Emotional movements are also lost in the area since the lesion responsible affects the common path of all the peripheral fibres.

In some lesions of the facial nerve the chorda tympani nerve may be involved. In this case there may be loss of taste on the anterior two-thirds of the tongue on the affected side. Since the chorda tympani is given off within the facial canal at the base of the skull, lesions occurring below this will not affect the sense of taste.

## Bell's palsy

Bell's palsy is the name given to an acute paralysis of the facial nerve. It is practically always unilateral and no obvious cause is found in the vast majority of cases. Depending on the severity of the condition, some or all of the muscles of expression on the affected side may be weakened or paralysed. The extent of the paralysis is easily seen if the patient is asked to smile, to close the eyes, and to furrow the brow. There may or may not be loss of taste sensation on the anterior lateral part of the tongue on the affected side depending on whether or not the chorda tympani nerve is involved. Some patients complain of severe pain in the area of the parotid gland or the ear, although this is generally only in the early stages of the condition.

The initial treatment for Bell's palsy is systemic steroids in high dosage—

60 mg prednisolone daily for 5 days, then rapidly reducing, is a reasonable regime. The rationale behind this is a little doubtful; reduction of oedema around the facial nerve is the most usual explanation of the action of the steroid. Some workers have doubted the efficiency of this form of treatment, others have suggested that it should be adopted only if spontaneous resolution does not take place. None the less, it is the authors' view that, unless there are significant contra-indications to steroid therapy, it should be started at the earliest opportunity after diagnosis. The possibility of long-term cosmetic deformity in an unresolved facial palsy is such as to necessitate the taking of all steps which might help the patient. It is important that an eye shade should be used to protect an eye which remains partially open. On a longer term basis Faradic stimulation of the facial nerve may be helpful in restoring residual weakness— hand held, patient-operated small nerve stimulators are now available for this purpose. Surgical treatment in the form of the placing of fascial slings and similar manoeuvres later in the course of the disease has proved, in general, to be of very limited value and is now rarely attempted. However, direct nerve anastomosis of nearby minor nerve fibres by microsurgical methods has been reasonably successful, the motor function being taken up by the anatomosed branch.

# Extra-pyramidal syndromes

Extra-pyramidal syndromes are neurological disorders which affect pathways other than the principal ones concerned with voluntary movement. They are characterized by abnormality of muscular action with tremors. Parkinson's disease is the best known condition of this kind.

It has recently become evident, however, that symptoms of a similar kind may occur following the use of certain drugs, especially in elderly patients. In particular, sedatives and tranquillizers of the phenothiazine group have been implicated. These patients may occasionally come to the dental surgeon for diagnosis since the diskinesic symptoms are often most evident in the muscles of the masticatory apparatus. Repetitive movements or tremors of the tongue of mandible which may be of a bizarre nature should be suspected as having a drug-induced origin in patients of this older age group. Treatment is by withdrawal of the offending drug, although resolution may take some time.

## Anaesthesia and paraesthesia

The onset of anaesthesia or paraesthesia, either slowly or suddenly, in the distribution of a nerve may have localized or generalized significance.

Local causes of anaesthesia in branches of the trigeminal nerve include direct trauma to the nerve (as in fractures or traumatic extractions), pressure on the nerve by a foreign body (such as a tooth root or root canal medicament), inflammatory changes in relation to the nerve (especially osteomyelitis) and neoplasms encroaching on the nerve. More generalized neurological conditions which may affect the trigeminal nerve in this way include multiple sclerosis and the neuropathy of mixed connective tissue disease (Chapter 8). Similar changes may occur as the result of a variety of intracranial space filling lesions.

Following a cerebrovascular accident there may be a range of neurological abnormalities affecting the face and mouth. Among the more difficult to deal with are the changes in sensation, proprioception, and muscular control which may make denture wearing very difficult for the affected patient.

## The temporomandibular joint

In considering the diagnosis of symptoms arising from the temporomandibular joint it should be kept in mind that there are two distinct sources of such symptoms. The first arises from the muscles, joint structures, and other associated tissues as a result of abnormal physical activity within the joint. The second is from pathological changes in the joint itself. In the first case, there are no primary physical signs within the joint and diagnosis must be made from the history, secondary signs, and symptoms. Abnormalities of movement and function may be of vital importance in giving the diagnosis in these circumstances. In the second of these cases there may be primary physical signs present in the joint to account for the symptoms. In such circumstances the joint damage may be associated with systemic abnormality—rheumatoid arthritis is an example of this situation.

Differentiation between these two groups of conditions may initially be very difficult. However, it is true that the very large majority of cases of pain and other symptoms arising from the temporomandibular joint do so as a result of a dysfunction syndrome rather than as a result of any primary pathology.

### Symptoms in joint disorders

1. *Pain.* The predominant complaint is of pain. This may be felt as a dull ache over the area of the joint, the ear, over the temporal fossa or over the maxilla. The pain may be bilateral or unilateral and is usually described as being constant, but with acute exacerbations. It is during these acute exacerbations that the radiation of the pain from the joint often occurs. In some instances associated pain in the neck, upper arm, occipital area, or along the lingual nerve may be reported. The severe attacks of pain occur predominantly in the early morning in some patients, whereas in other patients they are more common at the end of the day. Acute episodes may also be precipitated after a meal, at the wide opening of the mouth, or during the night when lying heavily on one side of the face.

2. *Joint sounds.* The patient most commonly complains of a click, representing the movement of one component of the joint over the others. This click may

be quite loud and readily audible. In other cases, however, the stethoscope must be used to hear the sound. Although an acute episode of pain may be precipitated by the action which also produces a click, patients find that pain is usually associated with periods in which clicking is minimal. Apart from the single loud click, crepitation-like sounds may be heard in the joint on stethoscopic examination. Care must be taken in the use of the stethoscope to eliminate the crackling sound produced by laying the bell of the instrument over hair; the resulting crackling sound can easily be mistaken for joint crepitation.

3. *Restriction in opening.* The patient may report difficulty in wide opening, often associated with the imminent onset of a loud click. In other instances the difficulty may be in applying pressure on closing the mouth.

4. *Swelling.* For some not very clear reason, some patients with temporomandibular joint disturbances complain of swelling over the maxilla. A slight degree of soft tissue swelling may be occasionally noted in this site on examination. In a few other instances patients may complain of tenderness and swelling in the area of the parotid, presumably an effect brought about by the close proximity of part of this gland to the temporomandibular joint. In these cases it may be a matter of some difficulty to distinguish between parotid involvement in joint disturbance or joint involvement in a parotid pathology.

## Examination of the joint

Examination of the temporomandibular joint and face should begin by observing the degree of symmetry of the mandible and face, and by observing the path of excursion of the mandible on opening and closing. Loud joint sounds may be heard at this time. In order to examine the joint by palpation the examiner should be in front of the patient so that movement of the mandible may be related to those palpated in the condylar heads. A single finger is placed over each condyle head whilst the mandibular movements are carried out. Abnormal tenderness associated with the joints is detected by light pressure over the fossa when the mouth is fully open. Faint joint sounds may be heard by using a stethoscope placed over the condyle head whilst mandibular movements are performed.

Following the extra-oral examination a careful note must be made of the occlusal relationship of the teeth, paying particular attention to missing teeth, the presence of faceting or any evidence of bruxism.

By far the most important abnormality of this kind in the present context is the absence of molar or premolar teeth, leading to a lack of posterior tooth support.

Muscular tenderness associated with joint disturbance may be detected by palpation of the masseters. This is carried out by asking the patients to clench the teeth firmly together. When the masseters are thereby put into contraction the examining finger is run up the anterior border intra-orally, counter pressure being exerted from the external surface. When the examining finger reaches the zygomatic origin of the masseter, tenderness becomes evident and is shown by the patient's reaction. A similar test should be carried out on the opposite side. This test for muscular tenderness is a more reliable one than that of the pterygoid sign. This procedure is misleadingly unpleasant to the normal patient.

Radiographic techniques used to investigate the joint include the lateral oblique view, used routinely to demonstrate the anatomy and range of opening of the joint (Fig. 11.1). The condyle necks and heads are demonstrated well in a high orthopantomograph view, (Fig. 11.2) although the relation of the condyle heads to the fossae is not well displayed. Tomography may be used to give a clear view of the condlyle *in situ* but the clearest view of the condyle head from the point of view of structural change is given by the Toller transpharyngeal view.

Left side

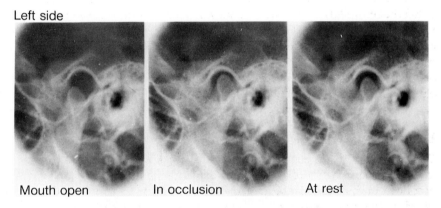

Mouth open      In occlusion      At rest

**Fig. 11.1.** Lateral oblique views of temporomandibular joint with mouth open, in occlusion, and at rest.

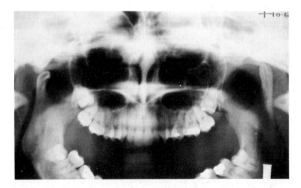

**Fig. 11.2.** High orthopantomogram showing abnormalities and assymetry of condyle heads and necks.

Arthrography, the injection of the joint spaces with radio-opaque material in order to delineate joint components is currently again being used after a period of disuse, particularly in association with tomography. Arthroscopy, the direct examination of the joint surfaces is now possible with the improvement in modern optical instruments although it cannot at the present time be considered a standard investigation. The place of computer assisted tomography (CAT scanning) and magnetic resonance imaging is also currently being assessed.

Electro-myographic studies of the musculature in cases of temporomandibular joint disturbance show deviations from the normal pattern of action. This is as yet a research tool, but it is probable that the technique will eventually be valuable as a diagnostic aid and much work is being carried out to this end.

## Chronic temporomandibular joint dysfunction (pain dysfunction syndrome)

This condition is found predominantly (80 per cent) in female patients. The predominant complaint is of pain, which may take any of the forms previously described. This pain may be associated with limitation of opening or with joint sounds, also as previously outlined. The patients quite often admit to a history of unusual mental stress although overt psychiatric abnormality is unusual.

There may be a history of previous joint clicking, limitation of opening, trauma, or recurrent dislocation. Examination may show one or several of the following: limitation of opening of the mouth; deviation of the mandible on opening; clicking heard or felt in the joint; gross malocclusion leading to abnormal joint movements or minor degrees of malocclusion with abnormal cuspal guidance of closure; gross occlusal attrition; occlusal disharmony resulting from the loss of teeth; poorly articulating dentures or bite closure resulting from lack of posterior teeth or unsatisfactory dentures; masseteric tenderness.

Questioning may reveal the presence of a habit of mandibular positioning or action, usually a protrusion or lateral excursion carried out when engaged on some mental activity. There may be a habit of biting on some foreign body such as a pen or pencil or there may be evidence of bruxism, supplied, in the absence of associated attrition, by some other member of the patient's family.

In most patients with chronic temporomandibular joint symptoms, radiographs of the joints reveal no abnormality of structure although limitation or increase in joint movement may be seen.

Care must be taken in differentiating the pain due to temporomandibular joint disturbances from that arising from dental causes or from facial pain of the types described above. In particular, the differential diagnosis between facial pain of psychogenic origin and that caused by joint dysfunction in patients undergoing mental stress is difficult. In fact, the two conditions occasionally seem to merge in a patient with physical signs of temporomandibular joint dysfunction, but with the demonstrative, anxious, obsessive outlook typical of the patient converting hidden anxieties into facial pain symptoms. Needless to say, in such cases, the first presumptive diagnosis must be of joint dysfunction and only if all physical signs of this have been cleared by treatment must the diagnosis of psychogenic pain be more firmly entertained. It is in observing the reaction to treatment that the differential diagnosis is perhaps best made.

## Rheumatoid arthritis

It is now accepted that the temporomandibular joint is involved to some extent in a large proportion of patients with generalized rheumatoid disease. In one survey 62 patients with established R. A. were examined, of whom 61 per cent had evidence of involvement of the temporomandibular joint. The major complaint was of pain followed by aching, stiffness, and crepitus. On examination, crepitus was found to be the most commonly occurring clinical sign, followed by tenderness of the joint and the existence of a functional abnormality. Radiographic evidence of changes was found in a high proportion of these patients, presenting as erosions, proliferations, and flattening of the condylar head. As in other joints, the disease process may occur in a phasic manner, acute exacerbations being followed by either a healing phase or a secondary chronic phase.

Immunological tests for rheumatoid arthritis have been outlined in Chapter 2. These tests may prove positive before systemic changes have been noticed.

## Osteo-arthritis (Osteoarthrosis)

Changes may take place in the temporomandibular joint as part of a generalized arthritic condition. In this case the history is usually of gradually increasing stiffness and pain. The joint may be tender to pressure and crepitations may be heard and felt when the joint is moved. Most patients are female and over the age of 50. The R. A. factor is negative and the E. S. R. normal in these patients.

Radiographically, the joint space is reduced and there may be some erosions of the articulating surfaces of the condyle and of the fossa.

## Headache and the TMJ

As will be clear from the above, one result of muscular pain associated with TMJ dysfunction may be headache. Similarly, pain arising within the joints themselves may be interpreted as earache; many patients found to have painful joints will have had an ear examination with negative results. The rather diffuse area between TMJ-induced pain and atypical facial pain has been mentioned. The few statistically based studies in this field seem to indicate considerable confusion in diagnostic criteria. Another field in which there is confusion is that of migraine. As has been pointed out, this is a condition with clear clinical criteria, an established genetic basis and with a known vascular and biochemical basis. Unfortunately, these criteria are often not applied before the diagnosis is made and there is no doubt at all that some patients with TMJ dysfunction are diagnosed as suffering from migraine.

The situation is made more complicated by the fact that, in some patients, there appears to be a triggering effect whereby one form of pain may be induced by a second mechanism. For example, some patients with fully diagnosed classical migraine have been shown to have concomitant signs of TMJ dysfunction. These patients may gain considerable relief from the migraine by treatment of the TMJ problem. None the less, controlled studies have shown that, in general, treatment as for TMJ dysfunction is ineffective in the management of uncomplicated true migraine.

## Management

A full discussion of the management of TMJ dysfunction is well outside the scope of the present book. The basis of management lies in accurate diagnosis, and the recognition and correction of aetiological factors. Restoration of occlusion, exercises, the use of occlusal splints, and various forms of physiotherapy may all play a part in treatment according to the precise nature of the diagnosis. Therapy with analgesic or relaxant drugs may have a limited use, but clearly, over a restricted period of time. As a last resort surgical techniques, including cryotherapy to the joint or greater auricular nerve may be employed.

# INDEX